The
Chinese Garden

History, Art and Architecture

THE
CHINESE GARDEN
HISTORY, ART AND ARCHITECTURE

Maggie Keswick

Revised by Alison Hardie

Contributions and conclusion by Charles Jencks

Harvard University Press
Cambridge, Massachusetts
2003

For John and Clare Keswick

What a delight to see Maggie's book mostly in colour and with a new Introduction, by Alison Hardie, setting it in context. When John Nicoll, our friend, came to me with these propositions there was not the slightest hesitation, and I'm sure Maggie would have been most gratified too. She died in 1995, but from the beginning it was always her intention to modify the book as interpretations changed and new material emerged. This notion of constant transformation was a lesson we learned together as she researched *The Chinese Garden* in the mid-1970s: the Tao, the Way of capturing the restless spirit of nature – several Chinese scholars emphasized – was more important for the garden than any of its particular motifs or themes. This book, dedicated to her parents, now in a different form continues its life for our children, John and Lily, who each have a special love for this unusual form of art and living.

Charles Jencks, 2002

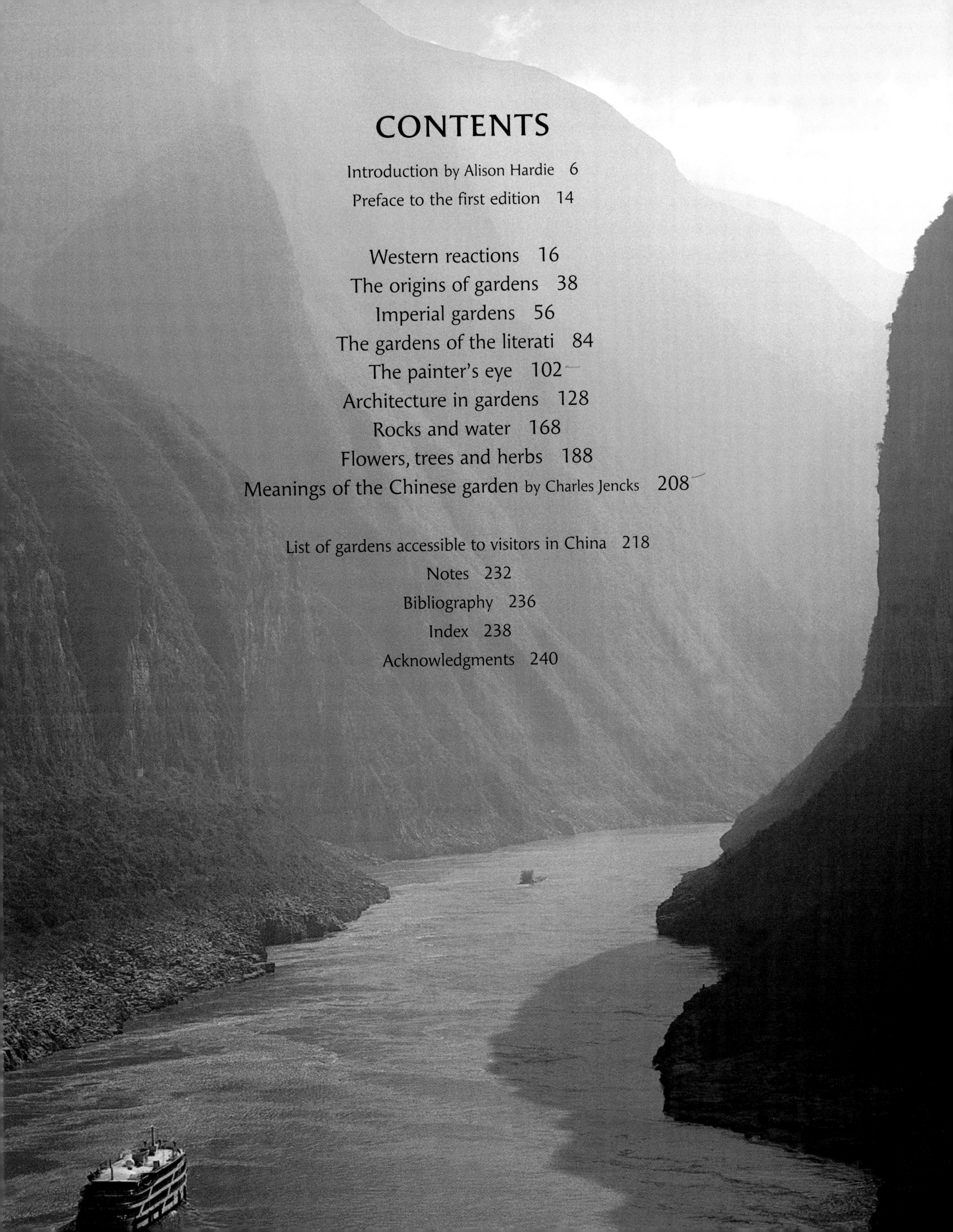

CONTENTS

Introduction by Alison Hardie   6

Preface to the first edition   14

Western reactions   16

The origins of gardens   38

Imperial gardens   56

The gardens of the literati   84

The painter's eye   102

Architecture in gardens   128

Rocks and water   168

Flowers, trees and herbs   188

Meanings of the Chinese garden by Charles Jencks   208

List of gardens accessible to visitors in China   218

Notes   232

Bibliography   236

Index   238

Acknowledgments   240

INTRODUCTION

by Alison Hardie

When Maggie Keswick (1941–1995) first published *The Chinese Garden: History, Art and Architecture* in 1978 (second edition 1986), there were very few introductions to Chinese garden history available to the general reader in English. The most authoritative was the Swedish art historian Osvald Sirén's *Gardens of China* (1949), with its companion volume, *China and Gardens of Europe of the Eighteenth Century* (1950).[1] Sirén's *Gardens of China*, to which Maggie Keswick pays tribute in her Preface, was the first authoritative book-length study of Chinese gardens published in English in the modern era; although no longer easy to come by, it is still of value, especially for its many excellent photographs, taken by Sirén in the 1920s and 1930s. Various other books and articles on the subject of Chinese gardens had been published in English in the first half of the twentieth century; although some were of real merit, they were often published in journals not readily accessible to the general reader in the West.[2]

A considerable amount of pioneering work in studying and recording Chinese gardens was done by Chinese scholars in the first half of the twentieth century, particularly under the auspices of the Chinese Architectural Society (Zhongguo yingzao xueshe).[3] The Society was responsible for making known one of the most important theoretical texts on Chinese gardens, the *Yuan Ye* by Ji Cheng (b. 1582). Originally published in about 1635, this was republished by the Society in 1933.[4] Sirén's book quotes extensively from this work, in a partial and unreliable translation, which seems to have been made for his use; it is Sirén's translation from which Maggie Keswick quotes extensively. At the same time, Tong Jun (Chuin Tung) was collecting the material which was to appear in his essential *Gazetteer of Jiangnan Gardens (Jiangnan yuanlin zhi)*,[5] to which Maggie Keswick refers. However, owing to the poverty and turmoil of China at this time, comparatively little was published, and the little that appeared was not easily accessible to the general reader, even assuming that he or she could read Chinese. It was not until the comparatively peaceful conditions which succeeded on the Communist victory in 1949 that scholarly work could be undertaken again with a reasonable prospect of publication. It was at this time, for example, that Chen Zhi and others worked on the annotated edition of Ji Cheng's *Yuan Ye*.[6] However, the upheaval of the Cultural Revolution, beginning in 1966, meant that much publication had to await the reform era after 1976, when many scholars had already met their deaths.

By the time that China opened up again to foreign tourists and visitors in the late 1970s, much of what had been written on Chinese gardens in English before the Communist revolution was out of date. Maggie Keswick was therefore something of a pioneer in introducing Chinese gardens to a wider, garden-loving audience in the West. Her background made her particularly suited to this task. She was the daughter (and only child) of Sir John and Lady Keswick; Sir John was taipan (Chairman) of the great Hong Kong conglomerate Jardine Matheson & Co. Ltd.,

The group of buildings known as the Revolving Archive, on Longevity Hill (Wan Shou Shan), the Summer Palace, Peking.

which had started out in the notorious nineteenth-century opium trade but had diversified into a vast range of business in East Asia.[7] Sir John lived much of his life in Shanghai and then in Hong Kong, spoke fluent, if hit-and-miss, Chinese (more Shanghainese than Mandarin), and had a real sympathy with Chinese ways of life and thought. Maggie inherited that sympathy, but her artistic gifts turned her more towards Chinese visual arts than Chinese business culture.

The image usually presented of the British community in pre-1949 China is that they lived lives cut off from the society around them, but the Keswicks had a large circle of Chinese friends, many of whom stayed on in China after 1949; after suffering for their capitalist backgrounds during the Cultural Revolution of 1966–76, they re-emerged as influential figures in China's reform era. So when Jardines as a company and Sir John Keswick personally re-established contacts with China in the 1970s, there was a ready-made circle of people who could facilitate all kinds of activities, including visits to the gardens of Suzhou. These were only just beginning to be restored after the ravages of the Cultural Revolution.

Maggie herself, after studying English at Oxford University, had become a clothing designer; she ran a successful boutique for some years in the London of the Swinging Sixties, before taking up the study of architecture. At this time she met and later married the influential architectural critic Charles Jencks. An able water-colourist and draughtswoman, she was also sensitive to contemporary trends in landscape and architectural design. It was no doubt her own warmth and charm, as well as her parents' contacts, which won her permission to visit many gardens in China which were barely if at all open to the public. Her sensitive response to their subtleties of design and her whole-hearted enjoyment of their pleasures come across very clearly in her writing.

Although she was always interested in design, and became fascinated by Chinese gardens when she regularly visited China with her parents from the 1970s onwards, it was not until later in her life that she became actively involved in garden design herself, both in the grounds of her family home in Dumfriesshire and working with the architect Frank Gehry in the USA. Her awareness of the psychological and emotional aspects of the garden environment came from her familiarity with the Chinese tradition and eventually fed into the charitable work which had always been part of her life, partly as an outcome of her mother's and her own Catholic faith. In the early 1980s she and her father had established the Keswick Foundation in Hong Kong, which started out by pioneering a number of pilot schemes in the area of mental illness, still at that time much stigmatized in the Hong Kong community. The Foundation later helped to set up a hospice for the terminally ill, the first of its kind in Hong Kong. After Maggie's first bout with cancer in 1988, she appreciated the importance to cancer patients and their families of having a place where they could draw strength from a pleasant and supportive environment. This led to the establishment of the first 'Maggie's Centre' at the Western General Hospital in Edinburgh, where she had been treated. Its design incorporated ideas about how to create a positive healing environment, partly drawn from the Chinese garden tradition.

After the publication of *The Chinese Garden*, Maggie Keswick continued to write and lecture on the subject. She wrote about Chinese gardens in architectural journals,[8] and 'An introduction to Chinese gardens' was published in *The Authentic Garden* (1991), the proceedings of a symposium on garden history which took place in Leiden in 1990.[9] She also contributed to a book published in 1990 to mark the opening of the Dr Sun Yat-sen Classical Chinese Garden in Vancouver.[10] Although somewhat reluctant to commit herself to paper, she was a lively speaker who did much to communicate her own enthusiasm for Chinese gardens to audiences in Europe and North America, even when she was already suffering from the cancer from which she died.

She and her book, *The Chinese Garden*, should be given a good deal of the credit for stimulating the remarkable growth of interest in Chinese garden history in the West in recent years. The subject has inevitably moved on since her book was published. Nevertheless, since other more recent studies have mostly been more specialized and thus less useful to the reader with little or no background knowledge, *The Chinese Garden* remains a very helpful general introduction. Given the very limited amount of material available to a non-Chinese-reading researcher at the time when she wrote, it is striking how balanced and well-informed her judgment is. To some extent her book appears to essentialize 'the Chinese garden' as an unchanging entity throughout the ages; these were the terms in which Chinese gardens had generally been discussed in the past. However, she was well aware of the constant cycle of destruction and reconstruction undergone by Chinese gardens, and does not fall into the trap of claiming that the gardens visible today are really survivals from the remote past. In fact, most 'classical' gardens which can be seen in China today are late nineteenth- or even early twentieth-century creations, which have moreover undergone extensive restoration (or reconstruction) since the late 1970s.

There has been a tendency among some Western writers to accept simplistic Chinese accounts of their garden history in an uncritical way. In particular, since the majority of Western writers on Chinese gardens (including Maggie Keswick herself) have been unable to read Chinese, they have not had the ability to consult primary sources, and some have seemed to accept the idea often put forward by Chinese writers on the subject that existing Chinese gardens on historic sites are essentially unchanged from their Tang, Song or Ming origins. This has led Western as well as Chinese writers to assume that the visual appearance of the Chinese garden has remained fairly constant throughout the ages, unlike the dramatic changes in style known to have taken place in European gardens. Maggie herself does not distinguish very clearly between the present and historical appearance of the gardens which she discusses, although she does recognize that dramatic changes took place in some instances.

There is also a tendency to privilege the gardens of Suzhou over all other Chinese gardens. Admittedly, this is a tendency which can be traced back several hundred years, but it is now particularly acute. There is very little popular recognition of the fact that many gardens existed and still exist beyond the Suzhou area, and that the Suzhou or Jiangnan style of private garden is only one of a number of regional styles. Aside from the imperial gardens of Peking, *The Chinese Garden* concentrates primarily on the gardens of Suzhou and other parts of the Jiangnan region, because these were the gardens most accessible at the time when Maggie was writing, but she was certainly aware of regional variations, and in later talks on Chinese gardens paid particular attention to the large-scale imperial park at Chengde (Jehol).[11]

A common misconception – not shared by Maggie herself – is that Chinese gardens were always urban and usually attached to the owner's residence.[12] In fact there is much evidence, particularly from the sixteenth and seventeenth centuries, to indicate that private gardens were more commonly situated in the part-inhabited, part-cultivated zones surrounding urban centres, outside the city walls, and were sometimes located in quite remote rural locations. Although gardens always included buildings, which made it possible for the owner and his family or friends to stay there overnight or for a more extended period of time, they were by no means always directly accessible from the family residence. The term 'private gardens' is in any case, as Maggie points out, something of a misnomer, since even privately owned gardens are known to have been open to visitors, at least at some periods of the year, from the Song dynasty onwards. The conventional image of the scholar recluse sensitively appreciating the subtle changes of the seasons alone in his elegant retreat should more realistically be replaced, as she recognizes, by visions of raucous drinking parties, elaborate banquets held to curry favour with high officials, and noisy groups of visitors enjoying gaudy lantern displays.

The distinction which is conventionally drawn, as it is in this book, between imperial, temple and private gardens is a somewhat artificial one, given the fluidity between these categories: the same site could alternate between private and temple ownership, and private individuals sometimes built garden residences in the immediate environs of a temple; private gardens could be taken into imperial ownership, and from at least the eighteenth century, if not much earlier, imperial gardens might partially imitate the style of private gardens; private gardens in the imperial period could also be of enormous extent, especially in rural areas, and in some cases must have rivalled imperial gardens in splendour.

An aspect of Chinese gardens which is much stressed in Western writing, and on which Maggie does lay great emphasis, is their spiritual dimension. Gardens are explained as microcosms of the natural world, as representations of the Three Islands of the Immortals, as incorporating the principles of *fengshui* (geomancy) and so on.[13] While this is unquestionably true, it is not an aspect that is much stressed in Chinese sources (partly because it goes without saying). Recent academic studies in the West have tended to focus on more down-to-earth aspects of Chinese gardens, such as their socio-economic significance, as I discuss below. This may be partly a result of an increasingly realistic attitude towards the mystic Orient, as well as of a change of focus in Western art history.

Garden history as an academic subject is a concept introduced into China from the West, probably through Japan as well as directly, during the period when Chinese intellectuals were reassessing their culture in the wake of the May Fourth Movement of 1919.[14] In China, gardens are conceptualized as a form of architecture (partly because Chinese gardens incorporate many more built structures than European gardens usually do), and it has been architectural historians who have led the modern study of Chinese gardens and their history. The pioneering work done by the Chinese scholars associated with the Chinese Architectural Society has been greatly supplemented since 1949, and especially since the end of the Cultural Revolution, by a wide variety of publications.

Initially, scholars tended to concentrate on individual gardens of historical interest, as can be seen from a group of articles published in the authoritative archaeological journal *Wenwu* (*Cultural Relics*) in 1957,[15] but subsequently more wide-ranging attempts have been made to examine the development of the Chinese garden throughout history,[16] to analyse Chinese garden aesthetics,[17] and to relate the characteristics of Chinese gardens to Chinese culture as a whole; an outstanding example of the latter is Wang Yi's massive study of the inter-relationship of gardens and the development of Chinese thought.[18] Many useful reference works have also been published, such as Zhang Jiaji's dictionary of Chinese garden culture,[19] and collections of sources on virtually all known historical gardens,[20] and specifically on those of Suzhou.[21] Essential to the student of Chinese gardens are the collections of essays by the distinguished scholars Chen Zhi and Chen Congzhou, many of which were written long before their publication.[22]

An area of study by Chinese scholars which has done much to illuminate the social aspects of Chinese gardens is research on early designers of gardens. Although historical gardens are often described as having been designed by their owners or by other famous literary or artistic figures, the practical work of planting, building and rockery construction had to be planned and carried out by expert craftsmen. The republication of Ji Cheng's *Yuan Ye* in the 1930s seems to have first stimulated interest in the lives of these craftsmen; such an interest also fitted with the exploration of Chinese popular culture following on the May Fourth Movement of 1919, and was given greater impetus by the founding of the People's Republic in 1949. Xie Guozhen's re-publication of three early biographies of the late Ming rockery designer Zhang Lian (Zhang Nanyuan, b.1587)[23] has been followed by detailed studies by Cao Xun and others of the lives and work of Zhang Lian, Ji Cheng and the eighteenth-century master Ge Yuliang.[24]

There have also been many publications in China and Taiwan which are primarily picture books, aimed at the tourist market and at Chinese, both mainland and overseas, wishing to rediscover the 'roots' of their culture. Indeed, in present-day China 'the Chinese garden' has become something of a symbol for the excellence of traditional Chinese culture. One of the reasons why the Chinese garden is such a potent symbol of China for the Chinese themselves must be the fact that it is one aspect of Chinese culture which has been widely admired in the West over several centuries, and which has actually had a significant influence on Western culture.[25] This makes it difficult for some Chinese people to consider the Chinese garden in a dispassionate way, and it is this emotional content which explains why the ways in which Chinese gardens are regarded sometimes strike the outsider as irrational. For example, as I have indicated above, there is a tendency to essentialize the garden and regard it as a constant in Chinese culture, although it is very easy to demonstrate that in fact 'the garden' has meant very different things at different times or places, or in different circumstances, during the course of Chinese culture, and that Chinese garden style has changed quite dramatically.

Since the end of the Cultural Revolution, ever-increasing cultural and academic exchanges between China and the West have stimulated the advance of Chinese garden studies around the world. Chinese garden studies in the West have generally become much more diversified than during the inter-war period, although the link with art history remains a close one. The incorporation of a reproduced Suzhou garden courtyard, the courtyard of the Late Spring Abode (Dian Chun Yi) from the Garden of the Master of the Fishing Nets (Wang Shi Yuan), into the Metropolitan Museum of Art in New York illustrates this connection.[26] A number of catalogues for exhibitions which have taken place in the United States have gathered useful material on and analysis of Chinese gardens, seen from an aesthetic and sometimes also a botanical point of view.[27]

Over all, since Maggie Keswick wrote *The Chinese Garden*, considerable new scholarly work has been done on various aspects of Chinese garden history. The study of Chinese gardens has been brought more into the fold of garden history as it is understood in the West with the publication of two special issues of *Studies in the History of Gardens and Designed Landscapes* (formerly the *Journal of Garden History*), under the editorship of John Dixon Hunt and the guest editorship of Stanislaus Fung.[28] Stan Fung, in particular, as a Western-trained architectural historian who also studied with Professor Chen Congzhou of Tongji University, has done much to bridge the gap between sinologists and garden historians.

Qing dynasty imperial gardens and parks have been studied in depth; in the case of the Yuan Ming Yuan by a group of scholars led by Michèle Pirazzoli-t'Serstevens,[29] supplemented by Régine Thiriez' study of photographs of the ruined garden taken by Western photographers working in China in the nineteenth century.[30] Philippe Forêt has studied the Qing emperors' summer retreat at Chengde (Jehol) from the point of view of its use in the Manchu imperial project.[31]

By one of those strange workings of the *zeitgeist*, several people, widely spaced geographically, started to work on the *Yuan Ye* from the 1980s onward: Stanislaus Fung in Australia, Che Bing Chiu in Paris and myself in (at that time) London.[32] Several new editions of the *Yuan Ye* – of varying merit – have also appeared in China in recent years.[33] The result is that a work which was known to readers of English at the time when Maggie Keswick's book was published only through the partial translation in Sirén's *Gardens of China* is now one of the best known and most thoroughly studied Chinese texts on gardens.

The extent of pre-modern and early modern Chinese writing on gardens has perhaps not been appreciated in the West, though this has been somewhat remedied by Craig Clunas' study of late-Ming connoisseurship literature in *Superfluous Things*,[34] and by articles which have appeared in *Studies in the History of Gardens and Designed Landscapes*: for example, Philip Hu on Mi Wanzhong's Dipper Garden[35] and Duncan Campbell on Qi Biaojia's Yu Shan Garden ('Allegory Mountain').[36] Robert Harrist's article in the same journal on garden names in Sima Guang's Garden of Solitary Enjoyment (Du Le Yuan) has also helped to illuminate the subtleties of classical Chinese garden nomenclature for English speakers,[37] as has John Makeham's later article on Confucian terminology in the garden.[38]

Our knowledge of the botany of Chinese gardens has been greatly expanded by the work of Georges Métailié in Paris on historical ethnobotany,[39] and of the Australian botanist Peter Valder on the garden (and other) flora of China.[40]

Maggie's fondness for the 'garden city' of Suzhou is clear from *The Chinese Garden*. The historical development of the city, including its many gardens, has been thoroughly researched by Xu Yinong, in his book on the development of the city's form.[41]

Craig Clunas, by arguing persuasively that the concept of the garden, at least in the Suzhou region, underwent a change from productive to aesthetic in the course of the Ming dynasty, has now made it impossible for scholars to discuss 'the Chinese garden' as an unchanging essence, or to accept unquestioningly that the Chinese garden did not change in appearance, at least during the Ming period, a time of dramatic change in other aspects of life and culture.[42] Still, no Western writer has closely examined evidence for the style or appearance of Chinese gardens at any given period (as opposed to the layout or other aspects of a particular garden), nor indeed have Chinese scholars done this in any great detail.[43] There is therefore still a tendency to assume that Chinese gardens have always looked much as they do now.

In addition, although writers on gardens acknowledge that in the modern period there are quite striking regional variations in garden style, this is usually attributed to the fact that existing gardens in Peking are almost all either imperial gardens (such as the Summer Palace [Yi He Yuan] or the Imperial Garden [Yu Hua Yuan] in the Forbidden City) or gardens belonging to imperial kinsmen (such as Prince Gong's garden), while surviving private gardens in Guangdong are acknowledged to be of very late date (generally late nineteenth century) and clearly influenced by European architecture.[44] Regional variations in style, therefore, are not looked for in the more distant past, and an unspoken assumption seems to be that the style of existing Jiangnan gardens (mostly private gardens in origin) is the style of all private gardens in the past.

Given the significant changes over time in other art forms, however, it seems intrinsically unlikely that the style of Chinese gardens did in fact remain constant over widely different times and places, especially during periods which saw rapid social, economic and cultural change. An important change of direction in Chinese garden design occurred in the cultural heartland of Jiangnan (Suzhou and Songjiang prefectures) around 1610–1620, and somewhat later elsewhere. This can be associated particularly with the famous garden designer Zhang Lian and also with his contemporary Ji Cheng; it derived from the changing attitudes towards landscape painting initiated by the very influential art critic Dong Qichang (1555–1636). The result was a change in taste from a more massive, overtly impressive and relatively axial style of garden to a more understated, naturalistic style.

Garden history in the West is by now a well-established field of academic study, but it is only in the last two or three decades that the garden has become accepted as a suitable object of study by social historians as well as art historians. From the late 1970s onwards, interdisciplinary research by scholars in the fields of English literature, geography and art history, such as James Turner, John Barrell and Denis Cosgrove, has changed our understanding of and attitude towards the uses of garden and landscape design and representation.[45] The insights of scholars such as these, relating changes in garden style and in attitudes towards garden culture to developments in society at large, have given us a new and deeper understanding of the many social as well as aesthetic meanings of designed landscapes and gardens in the European context, and are starting to prompt similar considerations in relation to non-European gardens. We are now less likely to see Chinese gardens, as Maggie Keswick did, primarily in aesthetic terms. But still, the study of Chinese gardens, in both Chinese and European languages, has remained more strictly within the realm of art and architectural history. In *The Chinese Garden*, Maggie does allude to the use of gardens as status markers, but does not analyse this use in depth.

In recent years, some Western scholars have started to apply to the history of Chinese gardens the insights of researchers in European garden history from a social science perspective. The social role of gardens was brought to the attention of sinologists in 1992, when Joanna Handlin Smith's important article on 'Gardens in Ch'i Piao-chia's Social World' was published in the *Journal of Asian Studies*.[46] Handlin Smith gives an acute analysis of the ways that gardens were used in late Ming Jiangnan both to express the owner's individuality and to situate him socially. The historian John Dardess was another pioneer in integrating the study of garden culture with Chinese social history, in an article and then a book on a particular county in Jiangxi province during the Ming dynasty.[47] A landmark came with the publication in 1996 of Craig Clunas' *Fruitful Sites: Garden Culture in Ming Dynasty China*;[48] this has become a seminal text for the understanding of the economic role of Chinese gardens and how they were used in a social context, and has moved the academic study of Chinese gardens in the West decisively away from the realm of aesthetics towards a socio-economic direction. Recent research by art historians of China has thrown light on the links between property ownership and the representation of landscape in Chinese painting, and this line of research shows signs of proving fruitful in the study of Chinese gardens.[49]

In this new edition of *The Chinese Garden*, the text of the book remains as Maggie Keswick wrote it, apart from some minor corrections of factual errors. Maggie was not an academic sinologist, and her romanization of Chinese names was erratic; all names (apart from those for which there are widely accepted equivalents in English such as Peking) have been converted to the Hanyu pinyin romanization system, which is now the most widely used. Some illustrations have been replaced with better or more up-to-date photographs. The appendix listing gardens in China accessible to visitors has been completely revised and updated. A list of garden names in English translation, pinyin romanization and simplified Chinese characters has been added, so that non-Chinese-speaking visitors to China can more easily find their way to the gardens, and enjoy the pleasures which so captivated Maggie Keswick and which she conveys so well in *The Chinese Garden*.

Newcastle upon Tyne, Summer–Autumn 2002

A solitary rose makes a splash of colour against an openwork window, as its shadow plays on a wall, in the Couple's Garden, Suzhou.

PREFACE

When people asked what I was working on, their reaction was nearly always the same: a book on Chinese gardens – who ever heard anything special about Chinese gardens? Even in the East they are something of a lost art; in the West the words seldom call up any image at all – or if they do, it is likely to be one of a Japanese garden, with its exquisite arrangements of stone and moss, its manicured pines and dry streams, and above all, its sense of being so perfect in itself that (as Mishima once wrote) even the intrusion of the visitor's own senses into the garden seems a violation.

Chinese gardens are not like this. Confusing and dense, dominated by huge rock piles and a great number of buildings all squeezed into innumerable, often very small spaces, for many foreigners the characteristic Chinese garden is so unlike anything else as to be incomprehensible and even, in parts, grotesque. Perhaps the closest analogy to the experience they offer would be something like a walk through Chartres cathedral. In both, the first, sensuous impressions lead on to more cerebral pleasures, and lying behind the forms, for those who wish to find them, are apparently unending layers of meaning which become increasingly esoteric and mystical as they are explored.

Like the plans of Gothic cathedrals, Chinese gardens are cosmic diagrams, revealing a profound and ancient view of the world, and of man's place in it. But in their long history they have also been, in quite a real way, the background for a civilization. In them China's great poets and painters have met and worked, and they have also been full of laughter and jokes, the scenes of ribald parties, amorous assignations, and the status-seeking efforts of countless *nouveaux riches.* As well as settings for peaceful contemplation they have seen family festivals and elaborate dramatics, political intrigues as well as domestic tiffs, in fact every kind of tragedy, joy and melodrama. For a woman with bound feet, her family garden might represent the sum total of her universe, and for some – like the maids in such novels as the *Dream of the Red Chamber* (*The Story of the Stone*) or Ba Jin's *Family* – the only way out of an intolerable existence was to steal away at night and slip into oblivion under the glassy surface of a garden pond.

A book on Chinese gardens is thus quite an undertaking, especially since so little has been written on them in the thirty-odd years since Sirén's classic *Gardens of China.* It seems a strange subject too, perhaps, in view of the momentous changes taking place in China. Yet a number of very old and beautiful gardens have survived wars and revolutions against all odds, and today are open to the public. Although the way of life that produced them has irrevocably vanished, they are immensely popular, nearly always crowded, and with increasing tourism they are once again accessible to the West.

This book grew out of many visits to these gardens made while my father was Chairman of the Sino-British Trade Council. In 1961 foreigners were still fairly unusual even in Peking, and with my mother we spent much time in the Summer Palace, and sometimes lunched in Beihai park, since it was then still open to the public. It was not until we travelled to Suzhou, however, that the gardens of China began to exert their extraordinary fascination, for there they are at their most beautiful and distinctive, and most clearly representative of ideas that are alien to Westerners. From then on I spent as much time as possible in Chinese gardens, and at home began to hunt through the libraries at S.O.A.S. and the British Museum for some light on their history and meaning. This book is the result of these searches. Inevitably many aspects of Chinese gardens have been left out. Apart from the historic gardens it seemed sensible to limit

the vast scope of such a subject by describing only those I had seen, and I have not visited any monastery gardens or many, many of the gardens mentioned in Tong Jun's (Chuin Tung) book of the 1930s. As it is I am deeply indebted to this book and also to Sirén's *Garden of China* of 1949 (including his translations from the *Yuan Ye* which I have used throughout). Invaluable too were a number of articles in learned periodicals, as well as chance references and insights in the works of many sinologists and art historians. A list of the main works used can be found in the bibliography, but I feel a special debt to the books of Edward Schafer, whose dazzling works on the Tang dynasty were a constant inspiration, and also to Andrew Plaks, whose analysis of the Chinese literary garden in the *Dream of the Red Chamber* crystallized and confirmed many ideas.

Without the impressive literature in English on many aspects of Chinese life, philosophy and art this book would have been impossible, but in addition I am especially grateful to the number of friends who worked with me as translators: particularly to P. D. Lu, Lillian Chin, Milly Yung, Kenneth Ma, Ken Shimamura, and Paul Clifford. The last of these also read the manuscript and regularized the modified Wade–Giles system which (except for place names written according to the accepted post office forms) was used for the romanization of Chinese words.

The line drawings were done from my photographs by David Penney, and the plans by John and Irene Corrigan. For pages 183 and 215 they worked from sketches adapted by me on site from the somewhat out-of-date plans in Tong Jun's book. For pages 78 and 80 they worked from sketches done in Peking by Frances Wood. I am also indebted to her and Agnes and H. W. Tang for bringing various works in Chinese to my attention.

Apart from all this, I would not have written this book without the help and encouragement of my parents and many friends, particularly Wan-go Weng, whose beautiful Qiu Ying scroll I illustrate, and on whose knowledge and kindness I constantly drew. I am equally grateful to Jessica Rawson, Roderick Whitfield and Y. C. Huang for their help and interest; to Wen Fong for the picture on pages 124–5; to Nelson Wu for giving up an afternoon to me in St. Louis; to Lawrence Sickman for an exceptionally enjoyable afternoon looking at scroll paintings in the Nelson Gallery, Kansas City, Missouri; and to the many men and women of the China Council for the Promotion of International Trade and of the China Travel Service, who acted as our guides.

I remain also forever in the debt of the late Lorraine Kuck, not only for her book on Japanese gardens which is a model for all historians of gardens, but for the most generous gift of translations by Grace Wan, together with many notes, which she had made towards writing a book on this subject herself. Once it was finished (and it was far too long), Bob Saxton was responsible for the many hours of judicious cutting and editing that shaped the book, and we also spent several days working together to choose the illustrations: to him and to Frank Russell, my editor at Academy Editions, I am also greatly indebted. Even with all this help, however, this book would not have been completed without Charles Jencks, who collaborated with me on the chapters on painting, architecture and rocks and water, and was mostly responsible for the last chapter. Without his help, enthusiasm and perseverance – in fact without his constant badgering and insistence – I should still be on the first chapter.

London, August 1977

Shadows play an important role in Chinese gardens. Wang Shi Yuan (Garden of the Master of the Fishing Nets), Suzhou.

WESTERN REACTIONS

LEFT *Detail of eaves, Imperial City, Peking.*

ABOVE *Versailles; nature arranged in perfect geometrical order by Le Nôtre.*

When the first descriptions of Chinese gardens reached Europe in the eighteenth century, they started a revolution in taste. Until then, ever since the Egyptians first laid out pleasure gardens on the banks of the Nile, the gardens of the West had been based on straight lines and rectangles. From Spain to Moghul India rich and cultivated men planted their decorative trees and flowers in formal patterns, bringing order and symmetry to nature, while from salons all over Europe the local princes looked down over meticulous arrangements that mirrored the organizing mind of man. Versailles, of course, was the most extreme example of this approach: at ground level the parterres might seem a little confusing, but from the high windows above all was revealed in perfect clarity with each pattern on the left balanced by its like on the right. Water, which in nature seeps and winds, was everywhere contained and either appeared fixed and static, like vast rectangular mirrors laid out on the landscape, or spouted from the mouths of mermaids and dolphins. The vistas of Versailles swept on into infinity as if they would encircle the earth, the clipped walls of the *allées* denying growth and change so that the Sun King might live in an illusion of absolute and timeless sovereignty.

Then, gradually, from the unimaginably distant palaces of the Chinese Emperor, intimations of a totally different approach to gardening began to filter through the courts of Europe.

The first complete description of a Chinese garden was published in Paris in 1749, in a letter from Père Attiret, one of several Jesuits whom the Qianlong Emperor employed as painters to his court in Peking. For several months each year these Jesuits were allocated a little studio inside the walls of an imperial retreat where, in summer, the Manchu rulers moved to escape the city heat. It was the Qianlong Emperor's favourite palace, a place where he could relax from official formality and indulge his greatest delight – the making of gardens.

For this Emperor landscaping was an obsessive passion. No sooner was one layout finished than he would start on another, and gradually several older Imperial gardens, as well as uncultivated and agricultural land, were absorbed into his domain. The result was a vast complex of lakes and palaces collectively known as the 'Garden of Perfect Brightness', or Yuan Ming Yuan. It was one of the most fantastic pleasure gardens ever built on earth.

An immense wall surrounded it. Outside stretched the dusty Peking plain, meticulously divided and ordered by generations of farmers. Inside there were at first formal courtyards and audience halls, but these gradually gave way to reveal, bit by bit, an apparently haphazard arrangement of hills, trees and water where the works of man, far from imposing any order on the landscape, seemed to have been designed expressly to fit in with its natural forms and contours.

For us today the idea of an irregularly planned garden is familiar enough. To Father Attiret it was a revelation. His letter describes how clear streams wound – seemingly as they willed – through gentle valleys hidden from each other by charming hills. On the slopes, as if by chance, plum and willow trees grew in profusion and through them paths meandered with the lie of the land

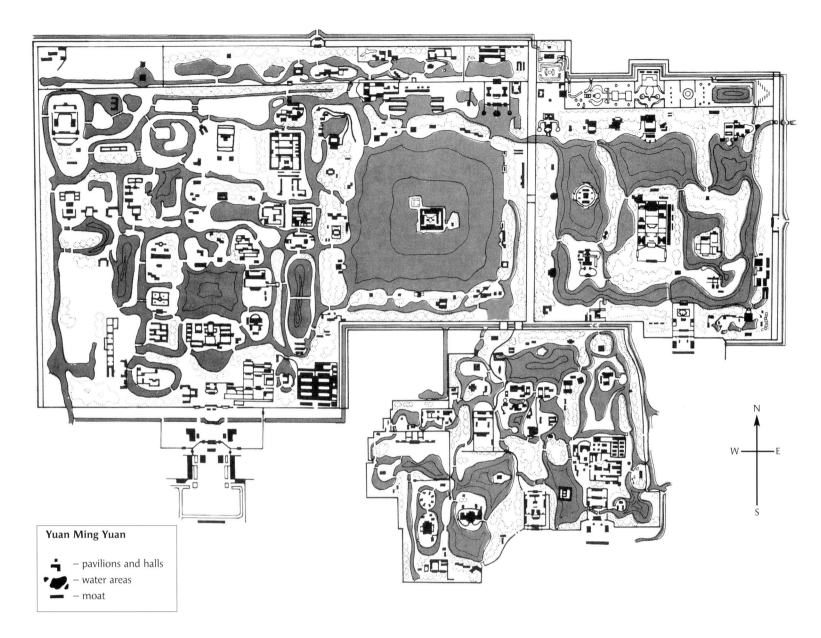

Yuan Ming Yuan

- pavilions and halls
- water areas
- moat

N
W — E
S

ornamented all along with little pavilions and grottoes. The streams themselves were edged with different pieces of rock, some jutting out, some receding, but 'plac'd with so much Art that you would take it to be the Work of Nature.'[1] For what was so intriguing was that the whole exquisite park was every bit as manmade as the landscapes of Le Nôtre: only in this case the Art of the whole endeavour lay in concealing, completely, any signs of the artificial. So, Father Attiret continues,

> in some parts the water is wide, in others narrow; here it serpentizes, and there spreads away, as if it was really pushed off by the Hills and Rocks. The Banks are sprinkled with Flowers; which rise up even thro' the hollows of the Rockwork, as if they had been produced there naturally.[2]

Each valley had its own pleasure house – each one different, each filled with its own collection of antiquities, books and objects of virtu. And linking these pavilions over the streams were zig-zag bridges which, if stretched out straight, would have measured over two hundred feet in length. On such bridges 'as afford the most engaging prospects' there were little gazebos, in which people could rest themselves, while, below them, the streams ran on until they met together to form wide lakes and waterways. One of these was nearly five miles round,

A plan of the Yuan Ming Yuan (Old Summer Palace) reveals it as a sinuous water labyrinth of lakes, winding streams, hilly islets and promontories. The whole complex was lavishly sprinkled with works of architecture, some forming formal courtyards, others dotted singly by the water or among the trees. Walls divided the garden into three main parts linked by water, and a boundary wall separated the whole place from the meticulous regularity of the farmlands outside.

A part of the Yuan Ming Yuan, near Peking, by Tang Dai and Shen Yuan (1744). This is one of the famous 'Forty views' painted for the Qianlong Emperor: hills enclose irregular valleys and winding streams ornamented with clusters of buildings – all different; some are grand and formal, others fanciful and decorative.

but what is the most charming Thing of all, is an Island or Rock in the middle of this Sea; rais'd, in a natural and rustic manner, about Six Foot above the Surface of the Water. On this Rock there is a little Palace, which, however, contains an hundred different Apartments. It has Four Fronts, and is built with inexpressible Beauty and Taste; the Sight of it strikes one with admiration. From it you have a View of all the Palaces, scattered at proper Distances round the Shores of this Sea; all the Hills, that terminate about it; all the Rivulets, which tend thither, either to discharge their Waters into it, or to receive them from it; all the Bridges, either at the Mouths or ends of these Rivulets; all the Pavilions, and Triumphal Arches, that adorn any of these Bridges; and all the Groves, that are planted to separate and screen the different Palaces.[3]

Symmetry and Disorder

Despite his enchantment with these scenes, however, it seems that Father Attiret anticipated a somewhat sceptical reaction from his friends at home in Paris. After all, he wrote, 'anyone who is just come from seeing the Buildings in *France* or *Italy* is apt to have but little Taste or Attention, for whatever he may meet with in other parts of the World.' He had felt like that himself – until he arrived in Peking. Yet there, despite all his training in the arts of European Order and

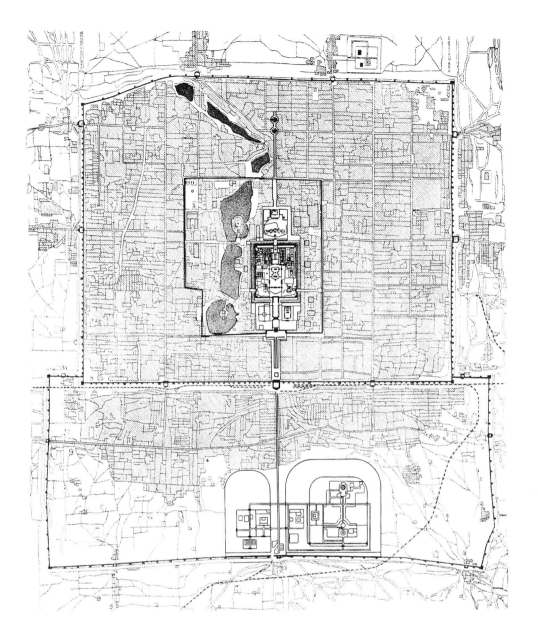

Plan of Peking. Within its massive walls, large main thoroughfares running north–south and east–west divided the old city of Peking into blocks of single-storey courtyard houses. In the centre, rusty red walls enclosed again the vast courtyards, ceremonial halls and living quarters of the Imperial City – all laid out in ordered sequence. But to the west of this the strict north–south axis is broken by the Jin Shui, or Golden Water stream, which flows diagonally across the city through a series of irregular lakes and parks. Although these great gardens thus seem to follow the natural meanders of the water, the whole thing is in fact entirely manmade.

Proportion, he found the palaces and pleasure house of the Emperor not only 'truly great and beautiful', but they struck him all the more forcibly 'because I had never seen anything that bore any Manner of resemblance to them in any Part of the World that I had been before.'[4] Indeed, he was obliged to admit that perhaps his 'eyes and ears' were 'grown a little Chinese', since, in works he might previously have found ridiculous, he now saw so much to admire.

The irregularity of the Emperor's gardens was not due to any lack of skill in the art of forming straight lines and right angles. In Peking Father Attiret noticed that the street plan of the city itself, the great Palace, the Courts of Justice, and the homes of 'the better sort of People' were all arranged with the strictest regard to geometric order.

For many commentators, the orderly succession of rooms and courtyards in Chinese domestic architecture have often been seen as an expression of the Chinese ideal of harmonious social relationships: formal, decorous, regular and clearly defined. Unlike many-storeyed European buildings, the houses of North China are arranged at ground level only, on a strict geometric grid, and instead of elaborate façades they present plain grey walls to the streets. In each courtyard the main room traditionally faces south with apartments of lesser importance placed

ABOVE, LEFT *Peking courtyards. View looking down on the traditional houses of north China in an old residential district of the city. The rectangular arrangement of houses facing inwards to private courtyards leaves blank walls facing the intervening alleys (hutong).*

ABOVE, RIGHT *Symmetry and regularity: the Imperial City, Peking, looking south from Coal Hill (Jing Shan) over the northern gate (Shen Wu Men). The layout of halls and courtyards is basically the same as in private dwellings, but on a vast scale. The double-tiered roofs are yellow-tiled, the great walls over which they brood a deep rusty pink. Much of the Imperial City as it now stands dates from the last (Qing) dynasty, but the layout and origin of the buildings are Ming.*

symmetrically on either side. In the old days, as a family's resources grew the courtyards proliferated, but they did not change in plan. In great establishments, as in small, the traditional arrangement of the courtyards and rooms helped to regulate and define the lives of those who lived in them.

The prescriptions for building houses thus allowed the builder little scope for personal expression, and the same was equally true of great official buildings. The Imperial City in Peking – incidentally the largest palace in the world – is in plan an immensely enlarged but still recognisable version of the ordinary Chinese house. Again, unlike a great European edifice, which makes its impression immediately, this palace has a cumulative effect. The first courtyard is entered through an arched tunnel in the massive outer wall. From there the visitor progresses through a succession of courtyards so immense that the central halls on each further side seem almost small in comparison with the horizontal expanse of the whole. In fact, all the buildings are raised above the ground on the high stone podium with which each courtyard is surrounded. Wide ramps slope up to the main halls. Above the podium walls and pillars are powdery red, and above that again, like horizontal strata in a cliff face, the shadowed eaves are painted in brilliant green and blue and gold. Symmetry reigns throughout. The central ramp to each main building is balanced on each side by two smaller ramps. Corner towers mirror each other exactly. A gate on the east has its twin on the west. There are two matching sets of golden water containers, and – the most powerful feature of the whole complex – the great clean-angled roofs, which rise and dip against the sky like gleaming yellow tents, are identically repeated on either side.

The effect of the Imperial City is one of perfect order perfectly imposed, an impression quite consistent with the absolutism of the Ming Emperors who built it. Looking around from the centre of each courtyard it is as if nothing inharmonious or irregular existed anywhere in the world, for above the great walls enclosing these vast spaces, only the sky is visible. The whole arrangement is one of extraordinary psychological force: not even at Versailles has symmetrical planning been used to such effect as an expression of total power.

It must have been obvious, then, to Father Attiret that the Chinese could apply laws of 'Order and Disposition' whenever it suited them. And clear too that the lack of this obvious order in their gardens was deliberate. Their principle, he wrote firmly, was to represent 'a natural and wild View of the Country; a rural Retirement, and not a Palace form'd according to the rules of Art'.[5]

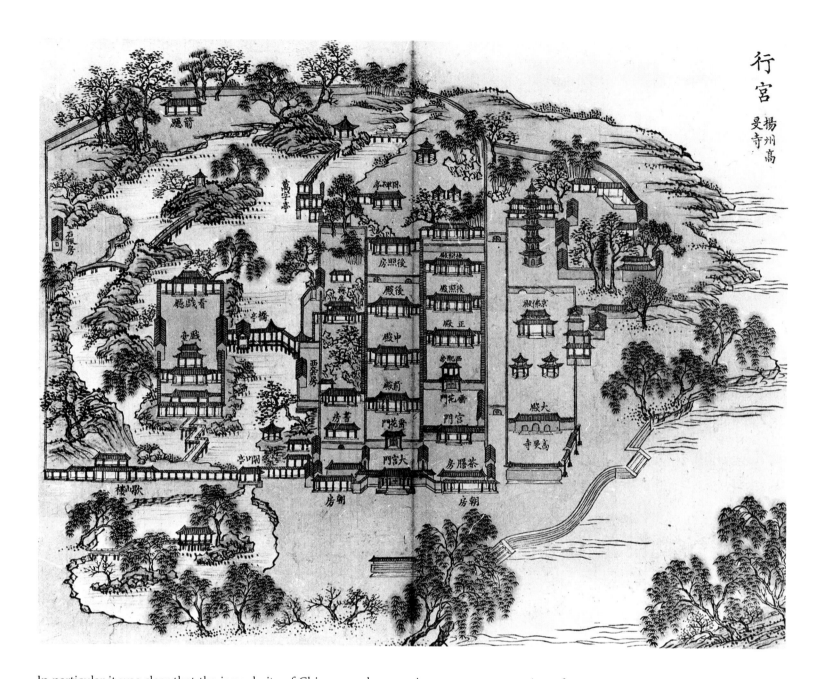

In particular it was clear that the irregularity of Chinese gardens was in no way an expression of disorder or of chaos. Their effect is, if anything, even more harmonious and a great deal less chilling than that of Versailles. There the formal grid marched out over the landscape as an expression of human supremacy while in the Yuan Ming Yuan it seemed as if man and nature were equal partners. In China straight lines and rectangles were reserved for artefacts which concerned man's relationship to man. When it came to man's relationship to nature, even the Emperor himself – whose domains far exceeded those of Louis the Sun King – did not presume to lay claim to sovereignty.

This special respect for the landscape had a philosophical basis. Side by side with the Confucian emphasis on man in society, a totally different view of the world, and how to live in it, had grown up in China. The people who had formulated its doctrines were known as Taoists, and they looked on man not as the measure of all things but as an inseparable part of the great Universe in which he existed. They sought to discover how this Universe worked, and removed themselves from involvement in worldly concerns. Rejecting the Confucian's fine distinctions, they believed in the

Bird's-eye view of a Buddhist monastery in Yangzhou, used as a temporary palace on imperial tours of the south. The rectangular courtyards of the ceremonial halls in the centre contrast sharply with the naturalistic gardens to the left.

fundamental unity of all things. To them, book learning and intellectual thought were inferior to receptive and intuitive knowledge. And rather than actively working towards improving the world, they believed that all would be well so long as things were left to run their natural course.

Taoism, so utterly opposed to the Confucian view, proved over the centuries to be its necessary complement, for it provided a release from the rites and duties of organized society. The Chinese (who are practical people) found both philosophies valuable and applied them side by side. Thus while the Chinese house mirrored the Confucian desire to regulate human society, the Chinese garden followed the Taoist principle of harmony with nature. And just as Taoism provided an acceptable escape from Confucian control, so the design of a garden released the creative imagination, which was allowed little play in the building of a house.

When Father Attiret felt obliged to explain his enthusiasm for Chinese gardens, however, he did so only on aesthetic grounds, claiming that 'All is in good taste.'[6]

The gardens had conveyed their own message to him, as they were meant to do, through the senses rather than the mind. Although he may not have understood their philosophical background, his descriptions still manage to suggest the richness of meanings inherent in the Chinese garden as well as something of their extraordinary magic – the sense, which is not lost upon further aquaintance, of being transported to a fairy landscape quite unlike any other on earth. As we shall see in the following chapters, this is not just a fanciful conceit.

English and Chinese Gardens

Attiret's letter became popular all over Europe and by the third quarter of the eighteenth century had been translated and read in all the fashionable salons. Of all those who enjoyed the account, however, it was the English who responded most sympathetically, for they had already begun to think along lines that seemed very similar to the Chinese approach. In Attiret they found their own gardening instincts supported from a delightful and unexpected quarter.

Back in 1692 Sir William Temple had already turned his readers' attention towards China in an essay called *Upon The Garden of Epicurus,*[7] in which he compared symmetrical European garden layouts with the spontaneous irregularity of Chinese examples. His term for this spontaneity was 'sharawadjii' which, since it does not correspond exactly to any Chinese phrase, has kept scholars busy ever since. Where did Temple find such a term? Did he make it up? And how did he hear about the Oriental approach to landscape design?

When, some fifty years later, the English read about the summer retreats of the Chinese Emperor, they were already well on the way towards their new style of gardening. Practice lagged a little behind theory, but in 1715 Stephen Switzer became the first professional gardener to make practical suggestions about implementing the new style. Shortly afterwards Alexander Pope moved into his villa at Twickenham and set about creating a garden where 'Spontaneous Beauties all around advance',[8] and no part of the land was touched without first consulting (in his famous phrase) the 'Genius of the Place'.

From then on snippets of information about the Chinese way of gardening were received in England with the greatest interest, but the great seventeenth-century Chinese manual on gardening, the *Yuan Ye,* remained totally unknown in the West. When Sir William Chambers wrote his *Discourse on Oriental Gardening* in 1772 his account was based on a brief personal visit to Canton in South China, and on hearsay – not on the reading of any Chinese works on the subject. In fact no English garden maker ever had any direct communication with a Chinese of similar interests, which is a pity for, when writing about landscape design, they often sound remarkably alike.

Chinese gardens were not only the products of Emperors. As early as the Han dynasty (206 BC–AD 220), as we shall see, rich men also liked to show off their wealth by landscaping their estates, and later, poets, painters, scholars and connoisseurs all added to the theory and practice of garden design. These men, superficially at least, were very similar in outlook to the men who made the new style of gardening in eighteenth-century England. Like them they were highly educated and members, on the whole, of literary elites. Some were rich, with the time and means to make over their estates in imitation of an idealized nature. Others, themselves scholars, painters and poets, worked for rich patrons as professional landscapists. They would have agreed with Pope that 'All landscape gardening is landscape painting'. Both emphasized variety in their designs. Stephen Switzer, for example, wrote that a gardener would 'endeavour to diversify his views, always striving that they may be so intermixed, as not to be all discovered at once',[9] while a typical Chinese garden is said to consist of 'more or less isolated sections which, though they succeed one another as parts of a homogeneous composition, must nevertheless be discovered gradually and enjoyed as the beholder continues his stroll'.[10]

When William Shenstone wrote about garden design in 1764 it was principally its appeal to the imagination that delighted him – something inherent in all the attitudes of the Chinese manual, *Yuan Ye*. This, too, has a whole chapter on 'Six Desirable Sites' for a garden, with poetic descriptions of how they may be enhanced, while in his *Unconnected Thoughts on Gardening* Shenstone is similarly interested in how to heighten and dramatize natural effects. He too has six categories under which to group the different qualities found in natural landscapes.

In particular, both Chinese and English gardeners drew inspiration from a classical antiquity, albeit in very different traditions. The Chinese liked to express literary allusions in the names of pavilions, and in poetic couplets which they hung on either sides of garden gateways: a reference to the 'Peach Blossom Spring' for instance, would have suggested many levels of meaning connected with an idealized vision of the past. In eighteenth-century England such allusions were more likely to be architectural: a Temple of Flora in the Classical style, carefully sited above the lake at Stourhead, suggested to the educated visitor a comparably lost world of pastoral harmony.

However, such parallels as these hide fundamental differences. For although it is easy, in selecting from works in both languages, to build up a body of principles that fit Chinese and English gardens equally well, in practice an English eighteenth-century gentleman would have felt totally lost in the garden of his Chinese counterpart. He would certainly have been surprised at the tight spaces, the many buildings grouped together in courtyards, and the lack of flowing, grassy distances. And he would have no doubt wondered at the intricate rockworks and grotesque standing stones, which would everywhere have commanded his attention.

The Wang Shi Yuan

Perhaps the easiest way to get an impression of this difference would be on a quick walk round one of the loveliest surviving gardens in China. Like many other gardens the Wang Shi Yuan is situated in a large provincial town, in this case Suzhou – a city of white-washed houses on a network

ABOVE *Francis Nicholson,* The Pantheon and Gothic Watch-Cottage *(watercolour, 1813), Stourhead, England. A Chinese scholar would have found the grassy spaces of the English 'natural' landscape garden somewhat empty and uninteresting.*

RIGHT *Plan of the Wang Shi Yuan, Suzhou.*

BELOW *Side canal, Suzhou. The city was founded in 500 BC by He Lu, who tamed this then barbarous region. It has been famous since early times for its villas and gardens. Long a place of political retreat, Suzhou is also celebrated for its many scholars and artists, among them Ji Cheng, author of the Ming dynasty garden manual, the* Yuan Ye.

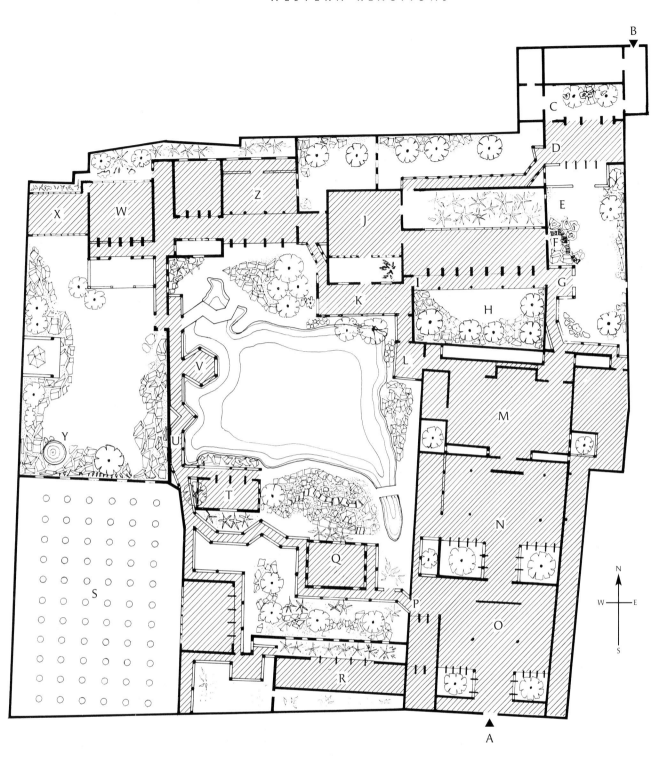

Wang Shi Yuan, Suzhou

— walls
— windows in walls
— walls below knee height
— walls with pillars supporting roofs
— roofed areas
— trees with shrubs
— flowers and shrubs in pots
— rock piles and hills
— bamboos

A South entrance
B North entrance
C Courtyard
D Hall (present-day shop) with terrace
E Courtyard
F Rock-pile with cavern and steps to top floor of Library
G Xie over entrance to library courtyard
H Library courtyard
I Door to gallery on lake
J 'Pavilion of the Accumulated Void'
K Gallery on lake
L Xie over entrance to house
M
N } Halls of house (two storeys)
O

P Entrance to garden from house
Q 'Barrier of Clouds' Hall
R Hall with secluded courtyards
S Area now used for pot culture
T Small hall on the water
U Covered galleries with calligraphy set into walls on stone tablets
V 'Pavilion of the Clouds and Moon'
W Hall with terrace
X Study
Y Well
Z 'Hall from which One Looks at the Pines and Contemplates the Paintings'

Total area covered is approx. 1 acre

of canals which has been celebrated since medieval times for its fine gardens. In the past other cities – Luoyang in Song times (AD 960–1126), Yangzhou in the late Qing (nineteenth century), Wuxi, Nanking, Hangzhou, Chengdu and Peking – were also much praised for their gardens. For the most part these have long since disappeared, but in Suzhou there are some twenty old private gardens still preserved, of which six have been restored and are now open to the public for a nominal fee. The early Spring is perhaps the best time to savour their particular charm.

The Wang Shi Yuan, meaning the 'Garden of the Master of the Fishing Nets', is one of the smallest of the Suzhou gardens. In contrast to the Yuan Ming Yuan with its seventy-mile circumference, it covers almost exactly one acre. It lies in the southern part of the city, well within the encircling walls, on a site which has been a garden since 1140. Unlike most other gardens still preserved, it is attached to a large house, now empty but open to visitors, which has double-storey rooms surrounding three courtyards.

Characteristically, the main entrance to the house (A on plan) is to the south, but at the north-east corner of the compound another gate allows a more direct access to the garden (B). It lies off a wide main street lined, in typical Suzhou fashion, with white painted houses and old plane trees. There, in the early mornings, brightly coloured bedding is hung out in the street to air. The doors into courtyards are all open revealing plump babies on every step, old men brushing their teeth, and

Garden entrance, Wang Shi Yuan, Suzhou. This alleyway leads off a busy street; the entrance to the garden lies through the unremarkable gate straight ahead, with the first courtyard set at right-angles to the entrance. No part of the inside is thus visible from the outside, and demons (or evil influences) are prevented from entering since they are unable to negotiate sharp turns.

other family bustle, while all around are sounds of daily greetings and bicycle bells. Alleyways sneak off between the exterior walls of the courtyard houses. The garden entrance to the Wang Shi Yuan lies down one of these.

The door is very ordinary. Once inside, the visitor finds himself in a simple white corridor open to the sky, twenty feet or so in length, with a blank wall at the far end. It is just like the alleyway outside, only painted white and more elegant in its proportions. Serving as a kind of decompression chamber, this passageway separates the noises of the city outside from the enclosed world of the garden towards which the visitor now progresses.

To the right of the blank end wall, framed in a high doorway, is a small and exquisite courtyard (C). Here a high white wall roofed with grey tile provides a backdrop for some strangely shaped stones, gracefully balanced on end, and softened by roses and a tall magnolia tree. On the left, facing the rock composition, the fourth wall of the courtyard is entirely made of pivoting full-length windows, finely carved and painted dusty red. They open into a shadowy hall (D). Beyond them, making up the opposite wall, similar windows lead out into the next courtyard (E). Shrubs, bamboos and some small trees are planted here as if the designer wished to introduce the growing elements of the garden gradually, without overdoing the greenery too soon. As in the first courtyard, the ground is paved with cobblestones, but here the paving is more elaborate, with a pattern of stylized flowers like a carpet.

The treatment of the ground sometimes disturbs Western visitors. In England there is a strongly developed cult of the lawn. Clipped and smooth, lawns define English suburban and cottage gardens just as surely as they surround the great houses of the past where, whenever possible, a ha-ha linked the lawn proper with the park beyond so that there was no apparent

RIGHT *Courtyard to the rear of the Late Spring Annex (Dian Chun Yi; W on plan), Wang Shi Yuan. In this picture the garden entrance shown in the previous picture lies to the right. Rocks and trees are framed by delicate carving against a plain white wall.*

RIGHT, BELOW *Doorway to courtyard, Wang Shi Yuan, Suzhou. The prospect of open space and colour beyond the featureless grey wall and plain surround of the doorway draws the visitor in. On the ground a floral design is worked in pale pink and grey pebbles.*

break in the long green sward. Today we can hardly conceive of a garden without grass. Yet Chinese gardens, as we see them today, have no lawns. The ground is either paved, often with fancy designs or, in more open or uneven places, left quite rocky.

Lawns, though we may not be consciously aware of it, have profound associations in the West. They are the formalized representation of grassy meadows, evocative of sweetness and light in nature, and behind them is the weight of the whole pastoral tradition. Nothing could be further from the traditional Chinese view. For the Chinese, soft grasses waving in the breeze suggested, if anything, the grasslands of the northern steppes and thus by association, the incessant raids made upon the frontiers of civilization by ferocious nomads whose lives were dictated by the need to pasture their cattle. A perceptive Chinese critic who visited England in the 1920s was amazed that any civilized person should want a 'mown and bordered lawn' which, he pointed out, 'while no doubt pleasing to a cow, could hardly engage the intellect of human beings'.[11]

Certainly, grass would be disastrous in a courtyard such as the one we have last looked at. The cobble designs are worked in soft greys and musty pinks, and they merge imperceptibly into the grey stones that form an irregular base around the enclosing walls. A light-green lawn would be much too prominent, defining the space too accurately and making the courtyard seem smaller. Indeed, the designer has made several subtle changes which increase the illusion of space; the long east wall, for example, is half-concealed behind trees and here is painted not white but grey, a soft, dark, cloudy grey which disappears like a misty shadow in the early light. There seems to be no wall at all, only a haze; and yet the area is ambiguous because the wall has a round doorway in it, promising more courtyards beyond.

The whole garden is in a sense a composition of courtyards. Some wind round corners out of sight. Others are half open-ended. Some are cut off like cul-de-sacs, or fit into each other like pieces of a puzzle. The total effect is of a labyrinth, with spaces layered round each other, again unlike, say, a French garden where space is as clear and distinct as French logic. This is partly why the extreme simplicity of the first impressions is so important. The blank purity of the alleyway entrance sharpens the visitor's responses in preparation for the various effects he will encounter in the following courtyard. The rocks and magnolia are in themselves intricate, but plainly presented against a white background. Standing in the cobbled forecourt, the visitor begins to focus on the details and juxtapositions of the arrangement and the changing effects of sun and shadow upon it. At seven in the morning the lower part of courtyard (C) is still dark, but high up the sunlight slants diagonally across the wall throwing shadows of the roof ends in regular waves on to the white surface. A little higher still, the magnolia blossoms open out in full sun; huge milky cups outlined against grey roof tiles. Their papery petals turn pink as they join the stem, like blood showing through sensitive skin.

Because such effects as these are so simple, yet so strikingly presented to him, the visitor begins to discriminate and slow down. It is a necessary step. As he penetrates further and further into the garden the effects will become increasingly complex. More and more rocks, walls, shadows, trees, buildings, roofs, pillars and patterns will all come crowding in, demanding attention. An almost unlimited series of opposite qualities will be played off against each other. Thus a dark narrow corridor between two high walls will be followed by a wide space full of sunlight and

Library courtyard, Wang Shi Yuan, Suzhou. The high walls of this viewing-garden with its decorative rocks and trees are broken by grillwork windows which allow glimpses of other courtyards beyond. The conical roof just visible over the wall belongs to the little pavilion on the lake (V on plan).

Carved trelliswork windows, Hall from which One Looks at the Pines and Contemplates the Paintings (Kan Song Du Hua Xuan), Wang Shi Yuan, Suzhou (Z on plan). This window looks northwards on to a small courtyard. The view of the lake to the south of the Hall is partly screened by rocks and pine trees.

leaves, or a plain, high wall by complex views of sweeping roof ends. Other contrasts are less obvious, like the treatment of space in the Wang Shi Yuan's second courtyard. Here, an enormous pile of convoluted grey rocks billows out into the enclosure like some kind of solidified cloud formation (F). The pile is fitted together leaving strangely shaped holes between individual stones to form an intricate pattern of contrasting solids and spaces – an effect much admired by sophisticated viewers (but quite alien to the untutored Westerner).

The purpose of the rocks, however, is not exclusively aesthetic. They also provide a necessary, if not entirely straightforward function. Winding through the pile is a series of narrow steps which lead up to a small door set high in the west wall. This door opens through the east side wall into the upper floor of a large library, which faces and backs on to two courtyards. This library has no inner staircase, so that the steps in the rock pile are the only way up to the first floor. It is a strange arrangement, for the ground floor cannot be entered from this courtyard at all, while the upper floor can be reached no other way. In fact the Chinese dislike seeing staircases and, as we shall see, often use such rock piles to disguise them.

Libraries and study rooms are an integral part of nearly all Chinese gardens. Their owners' time

was often spent writing poetry and practising calligraphy, often in the company of friends, and no garden worth its trees lacked a suitable study pavilion. Such rooms were usually secluded and surrounded by their own private courtyards to protect the reader and give him a pleasant prospect on which to look out. In the Wang Shi Yuan the floor of the library is raised on a stone podium, with a view over shrubs and rocks from behind large brown-painted pillars. The exterior doors, which pivot so that the whole of the interior can be opened up, are grand and formal. The adjacent courtyard (H) is very enclosed, surrounded by high walls, and its arrangement of shrubs and stones is peaceful, pleasant and deliberately not too dramatic. In case it should seem claustrophobic, a tall *Sophora japonica* tree has been allowed to grow in the furthest corner, its leaves rising high over the restraining wall. Beyond – as if the tree had been specifically planted there to draw the visitor's attention in this direction – the topmost portion of a pointed pavilion roof (V) is visible over the courtyard wall.

Such glimpses keep us moving on through a Chinese garden. It is as if the designer were constantly holding back, enclosing each view, yet always suggesting new delights just beyond the further wall. Indeed, the succession of different courtyards has been building up an increasing sense of anticipation as our visitor begins to feel, however unconsciously, the rhythm of the design. At one level the succession of contrasts has worked quickly: the narrow entranceway leading to a courtyard, the dark hall leading into sunlight, the grey tiles set off against white walls. However, the quick contrasts have also set up a longer-term expectation, the feeling that sooner or later these small-scale juxtapositions will broaden out into something quite different, something more comprehensive.

It is, therefore, exciting but not altogether unexpected when through a small door in the end of the library wall the visitor begins to see a different kind of light, a moving dappled light that plays on the walls beyond in a way that has not happened before. It is, of course, a reflection from the level surface of a lake – the smooth expanse of water which the visitor, lingering among stones, walls and cobble walks, has subconsciously been led to expect.

ABOVE, LEFT AND RIGHT *Interior courtyards of the house, Wang Shi Yuan, Suzhou. The three ground-floor halls (M, N and O on plan), are lit only by light spilling down through these little interior courts, called 'heaven's wells', since they also catch the rainwater from the roofs. The elegantly carved doorway is the main entrance to the house. The fine, paper-covered windows pivot from floor to ceiling. The garden lies hidden beyond the plain white wall to the right.*

RIGHT, ABOVE *North-western courtyard, Wang Shi Yuan, Suzhou (W on the plan). View looking south from the garden's most secluded room. A courtyard for pot culture lies beyond the far wall. The rocks to the left conceal a deep pool which seems to well up from a spring within the rocks.*

RIGHT *Western boundary wall of the Wang Shi Yuan, Suzhou. This attached pavilion contains a fine standing rock from Lake Tai near Suzhou. Seats have been set between the pavilion's side pillars, so visitors can contemplate the rock under cover. On the extreme left, stone steps lead down to a rocky pool, so arranged that the water seems to well up into it from some cavernous depths below.*

It is this part of the garden that is really its heart. The fairly small pond, roughly edged with rock, is made to seem larger because its irregular shape cannot be seen all at once. Around it are grouped halls and summer houses, some set back behind rocks or terraces, some right at the water's edge. Across from the wide, roofed entrance gallery (K), which the visitor reaches through a narrow doorway (I) from the preceding courtyard, stands the little pavilion (V) whose roof appeared above the courtyard wall. Apparently floating out over the lake, its stilts shadowy and half-invisible among the rocks, the pavilion is the focal point for the whole little lake. Setting off south round the water towards it, however, the visitor will find himself diverted from his purpose by many other charming resting places, unexpected groves and little works of architecture. There is an elegant hall (Q), for instance, which is totally screened from the lake by a large manmade mountain of rocks and earth. It is dark and shadowy inside, with a fine fretwork window framing a view of sunlit rocks. Beyond it, to the south, is a secluded bamboo grove, and in the wall which frames it are grillwork windows. Through these, half-screened and blurred among the shadows, are more trees and another grey-tiled roof (R).

By the time our visitor finally gets to the pavilion over the water, he has seen it from several different vantage points, while often it has been completely hidden. On the way he has passed no fewer than five other buildings, each with its own particular atmosphere: a density of experience that is quite overwhelming if you are used to the slower pace of western gardens. He may even have entered the side door of the private apartments. Here he would have walked through the large, dark halls of the house itself (M, N, O), finally reappearing in the garden at a totally new point (P), from which angular galleries set off through a bamboo grove. Behind one wall to the south-west is the area of the garden set aside for pot culture (S), with its numerous dwarfed trees and herbaceous plants set out in rows of earthenware pots.

Once reached, the pavilion is a delightful resting place. As in all such pavilions, low seats are built in between the supporting pillars, and the balustrades are also backrests, curving out over the water to receive cool air rising from the lake with, above, a view of reflected light playing on the inside of the eaves. Below, the water spreads out peacefully, reflecting rocks, overhanging trees and other pavilions in its milky surface; and goldfish gleam from murky depths, suggesting another dimension beneath the sunlight. Across the water can be seen the doorway through which the lake was first sighted, the entrance gallery, and the portico to the private apartments (L). Along a covered gallery (U) behind the pavilion, many grey stone slabs have been let into the wall, engraved with the poems and memories of those who have sat here in the past. Again, to Westerners, this is an unexpected convention, while for the Chinese an old garden which does not include calligraphy is in some sense unfinished: its presence adds significance and beauty to the charm of ancient trees and stones.

The visitor may be surprised to discover at this point that, although the Wang Shi Yuan is a small garden, he is now only halfway round it. To the north there is a whole collection of intersecting buildings, with rocks, walls and courtyards dividing and linking them. Some are private and intimate, some more public and exposed. A zig-zag stone bridge wriggles across a narrow channel to a small hill where tall pines are planted. Behind it is a fine hall (Z) for entertaining guests. At the north-west corner the visitor will eventually come across what is perhaps the most secluded of all the garden buildings: a tiny room (X) set slightly behind and to the side of a larger hall (W). Here he will at last feel that he has reached the furthest point. There is no way on, only back – through the main hall, along a gallery, left through the 'Pavilion of the Accumulated Void' [sic] (J), out into the courtyards to the north of the library, and finally back to the entrance, the narrow white alley leading to the world outside.

RIGHT *Wang Shi Yuan, Suzhou, looking east along the bridge running in front of the Hall from which One Looks at the Pines and Contemplates the Paintings, towards the pavilion at the entrance to the house (L on plan).*

PAGES 34–5 *Part of the Qianlong Emperor's great eighteenth-century garden, the Yuan Ming Yuan (Old Summer Palace), near Peking, painted in 1744 by Tang Dai and Shen Yuan. Note the rectangular gallery raised above the lake on stilts, the numerous pavilions dotted about the shores, and the use of individual, vertical rocks at the mouth of the stream to the right.*

The Ji Chang Yuan, Wuxi (above), dates from the sixteenth century and, although these pavilions are much later restorations, its layout remains as it was. In the eighteenth century the Qianlong Emperor, inspired by a visit to this garden during one of his southern tours, commissioned the Xie Qu Yuan (below), in the Yi He Yuan (Summer Palace), near Peking. The plans (see pages 80–81), are faintly reminiscent of each other, but the softness and gaiety of the southern garden has given way to a more formal elegance in the north.

At the end, the visitor will have no more idea of the plan of the garden than he did when he started. He will feel that there are innumerable parts he never had time to explore. Certainly he will find it hard to believe that the garden, and the house, are totally contained inside an acre of ground. And if he is used to the flowing lawns and country spaces of the *jardin anglo-chinois* (as the French like to call this English invention), he will marvel that anyone could have associated them with China.

When the excitement of Father Attiret's letter had died down, it was clear that Europe, and England with it, had bypassed the reality of Chinese gardens, and interpreted them entirely in the image of their own experience. Although the West appropriated the principles of concealment and surprise, the winding waterways and paths, and some prettified versions of Chinese pavilions, all the strength and intricacy of Chinese garden design drained away in the course of transference to European soil.

In 1772 Oliver Goldsmith wrote a satirical piece on the current vogue for things Chinese.[12] In it, a visiting mandarin is astounded to find a *chinoiserie* summerhouse described as a piece of 'Chinese' architecture. A hundred years later the same gentleman might have found it equally hard to believe that many plants then growing in English gardens had originated in China. And these had far greater claim to authenticity than the summerhouse.

In the nineteenth century the gardening interests of the West had turned more and more towards the display of rare and exotic plants. In China Western plant collectors discovered a paradise of flowers. However, the supply of new species from Chinese nursery gardens soon ran out, and the borders and woods of England and America gradually began to blossom with species of azalea and rhododendron found in wild foothills of the Himalayas. Many of these had never been cultivated in the gardens of China. Indeed, in the end, obscure hills and valleys in Yunnan and Sichuan did more to transform the horticultural traditions of the West than the designs of all the great pleasure gardens of China – and their underlying philosophy – put together.

THE ORIGINS OF GARDENS

The civilization of China took shape on a great plain which lies roughly on a level with the American state of Tennessee and the Mediterranean island of Crete. To the west the Huang He, or Yellow River, rises in the foothills of the Himalayas and flows east in a great loop to the seas that edge the Pacific Ocean. In winter this plain is cold and dusty, and the summer rains are erratic and precarious. It was never exactly a Garden of Eden. On the other hand it was by no means a desert. It was here, around 1300 BC, that a number of people began to merge and follow that common course which would eventually develop into Chinese civilization.

By the first historical Chinese dynasty (the Shang, 1600–1038 BC), these people had already settled down to agriculture. At their disposal was the plain itself, covered in a thick soil of loess and alluvial deposits. To the west stretched a vast terrain of yellow earth, cut into terraces by erosion, and hills thickly covered with primeval forests which in richness and variety have never been equalled. Here also grew many indigenous grains, fruits such as peach, pear, plum, persimmon and apricot, and herbs like the little yellow chrysanthemum, valued for its medicinal properties long before it was grown as a decorative plant.

Since they depended on agriculture, the early Chinese paid homage to the soil in what has been called 'a religion of agrarian fertility'. In some districts the life-giving rain was thought to be the seed of the Supreme Ruler (later called Shang Di), a celestial god who begot not only man but all living things on the body of the mother goddess earth. Man was not exalted as essentially different from all other created things, and his prosperity and happiness were seen to depend, like all forms of life, on his successful adjustment to natural forces. Such a view is common to many agrarian peoples, but in China it grew into an ideal of harmonious co-operation which became central to philosophy for more than two thousand years.

LEFT *Reed Flute Cave, Guilin. Many of Guilin's needle-pointed mountains contain vast, naturally formed limestone caves. One, decorated with stalagmites formed in groups like organ pipes, extends more than a mile into the hill. Some of these caves are new discoveries, some have been famous since the Tang dynasty. Popular legends linked such underworlds with the lands of the Immortals. Many artificial mountains contain hollowed-out rooms or 'stone houses' which, in addition to their magical suggestiveness, were places of refuge from the intense and cloying summer heat.*

ABOVE *Paddy fields at Yangshuo, near Guilin, Guangxi.*

Another later creation myth gives an illuminating picture of how modestly the Chinese regarded their own position in nature. According to this the origin of the world lay in a primordial egg which hatched a god who lived eighteen thousand years. Then he died. His head split and became the sun and moon, his blood the rivers and seas, his hair the plants, his limbs the mountains, his voice the thunder, his perspiration the rain, his breath the wind and his *fleas* the ancestors of man.[1] In fact, however, it is an idea that looks two ways at once, for although among the mountains and waters of the earth men may be no bigger than fleas, nature itself is seen to be formed in the image of a man.

It was an image which the Chinese did not hesitate to improve. Their aim of harmonious adjustment to the natural world did not make them submit passively to life. In fact they felt that man was the agent through which the abundant potential of nature could be fully realized. Thus

其most important myths idealized the first five Emperors from whom, so the legends went, the Chinese learnt the secrets of fire, animal husbandry, agriculture, irrigation and flood control. Armed with this knowledge and the assurance of divine instruction, they spread out over the landscape, cutting, terracing, digging, planting and watering until the forms and even the climate of north China would have been unrecognizable to their ancestors. Historians commonly agree that the Chinese have had more effect on their environment than any other people on earth before modern times, including even the ancient Egyptians, and nature repaid their diligence and devotion by giving them (at least until the middle of the eighteenth century) the most productive agricultural areas in the world.

Since it was all so successful, the Chinese could happily regard the alteration of their environment as an adornment rather than subjugation of nature. Indeed, numerous visitors have seen the intricate patterns of Chinese cultivation as a fitting symbol for the harmonious marriage of the Chinese with their land. Many have said that the tended fields make the whole landscape look just like a garden. If so, however, it is a Western, not a Chinese garden. For, unlike the Egyptians, however beautiful the mosaic ribbons of their dykes and fertile fields, the Chinese did not choose to reproduce these patterns in their pleasure gardens.

The Egyptians populated an inhospitable landscape and, when they came to make gardens, they had no model of luxuriant growth other than the one they had created themselves from the desert, so they planted their flowers in the same rational patterns as they had their crops. The

Leaning on the balustrade. Illustration of the Qing dynasty novel, Dream of the Red Chamber (The Story of the Stone). *Scarlet balustrades like this also act as back rests in all traditional Chinese gardens. When the gardens were privately owned open pavilions were often provided with summer curtains like these, for shade and privacy.*

Balustrade overlooking the lake, Wang Shi Yuan, Suzhou. The simple design of the balustrade is complemented by more fanciful carving under the eaves which, together with the vertical posts, forms frames for the different views.

walls round Egyptian gardens acted to keep out untouched nature so that the inside could be made beautiful by the hand of man.

The Chinese, on the other hand, were surrounded by a landscape of exceptional beauty and grandeur which existed beyond their fields and rectangular houses, independently of all their work and effort. This not only inspired them with awe, but reminded them of that great and impersonal order of the universe into which man, however successful and prosperous, was still obliged to fit. It was this image their pleasure gardens aimed to reproduce: thus the walls around a Chinese garden acted to block out the surrounding patterns of human activity so that the inside could be turned back again to nature.

Such an effect, however, could hardly develop before agriculture was well established, so the first reference to a 'garden' in Chinese literature suggests that it consisted solely of useful trees planted within a homestead's protective wall. This garden, containing willows, hardwoods and mulberries, is mentioned in the earliest work of Chinese literature, the *Shi Jing*, the 'Book of Songs' or 'Book of Odes' collected together by the sixth century BC, although the individual poems may date from considerably earlier. The same work also suggests a possible origin for great landscape projects in the ritual terraces built for such early heroes as the great King Wen of Zhou. The pits left by the excavations for these ancient terraces perhaps became the first of a series of artificially created imperial lakes that continues even today in the Parks of the Sea Palaces in Peking.

The First Pleasure Garden

One early work, the *Yi Jing* ('Book of Change') already links gardens with wild rather than cultivated nature when it speaks of the pleasure to be found in 'hills and gardens'. However, the earliest description of a true pleasure garden occurs in a poem of perhaps the fourth century BC, written in the southern state of Chu and included in the *Chu Ci*, or 'Songs of Chu'. It is specifically connected with the shamanist religion that played a powerful role in this and other early royal courts and, although it is an imaginary garden, the imagery of the poem suggests a highly developed tradition of garden design lying behind it.

Because of its context this early poem tenuously associates gardens with the magical cults of antiquity. Shamans and shamankas (that is, female shamans) played an important part in ancient Chinese life. By techniques of ecstasy they were able to leave their physical bodies and journey to amorous assignations with supernatural beings. Often they developed a special relationship with one particular deity whom they could persuade to look favourably on human requests. By association the priests of shaman cults were often half-divine themselves, and it is in their soaring, trance-inducing chants that the distinctive Chinese garden makes its first appearance.

The shamans are trying to heal a dying king by persuading his soul to return to his body. To do this they describe the lightfooted princesses who are waiting for him 'in your garden pavilion, by the long bed curtains'.[2] In the song, the king's palace is made up of shady chambers and high walls which lead off from tiered balconies; it has doors of scarlet latticework and rafters painted with dragons and serpents. In winter the rooms are warm, but in summer they all open into a garden that is described with such simple, conventional, but telling details that the dream starts to take on the feeling of an idealized reality. In the garden, according to the shamans' chant, the king will walk along linked galleries which catch the scent of orchids on the breeze. Streams meander past peacocks and a hedge of hibiscus flowers by the hall. There are loggias and covered walks, 'for exercising the beasts', pavilions tower over the palace roofs, and seated in the hall, leaning on the balustrade, you look down on a winding pool. Its lotuses have just opened; among them grow water chestnuts, and purple-stemmed water mallows enamel the green wave's surface.[3] Above it all are high, stepped terraces, from which the king can look down across the garden and up to the distant hills.

Perhaps this lovely place was as much a work of the imagination as was the perfect demeanour of the palace maidens, but it is strange how the description seems to stretch out across time like a premonition of what is to come. Already, in the shamans' magic chants, are the winding streams, the lotus lake, the balconied pavilion leaning out over the water, the scarlet fretwork, the terrace, and the shady galleries of the characteristically Chinese garden. Only the mountains are missing. For though the king can climb his terrace and look out at them from there, the distant hills remain firmly outside the garden walls; the rock piles, artificial slopes and sculptural stones that are to become the most arresting elements of a Chinese garden are so far nowhere to be found.

The Imperial Hunts

At the end of the Zhou period (1066–221 BC) the state of Chu was destroyed, and with it the elegant and sophisticated southern culture that had produced not only the *Chu Ci* poems but perhaps also the first Chinese gardens, as evoked for us in the shaman chants. After this destruction the development of gardens moved to the centre of power in the north, where life was tougher and less luxurious. There, instead of the private pleasance of the king, we find descriptions of vast hunting parks, where the hunts themselves became symbols of imperial might.

Detail of a hunting scene, drawing taken from wall painting in the Tang dynasty tomb of Li Xian, Prince Zhanghuai, in Qianxian county, Shaanxi. This tomb was excavated in 1971 and 1972. From ancient times, large tracts of land were enclosed as imperial hunting grounds. In time they were stocked with beasts and birds, and planted with rare and exotic plants sent as tribute from all corners of the empire.

The ancient oracle bones of China mention hunting grounds in present-day Shandong which were set apart for the Zhou princes. The *Shi Jing* too has poems which describe hunting parties, a pastime reserved, as in many other societies, for aristocrats and not part of ordinary people's lives. War and hunting were closely connected, the latter often organized as military exercises, deploying hundreds of beaters, foot soldiers, archers and horsemen. In the Han dynasty poets wrote magnificent descriptions of such imperial hunts which describe how the beasts were surrounded, driven, cornered and slaughtered with all the skill of a major battle. Mad with terror from thunderous shouts and flashes, birds and animals dashed themselves in frenzy against the chariot wheels: 'Hills and pools were obliterated in the process. Then the slaughter began with arrows and spears and swords wielded by the soldiers and their officers, a wind of feathers and a rain of blood covering the ground and darkening the sky'. Afterwards, so the poets said, 'the stiff corpses of the slain, bird and beast, glistened like tide-washed pebbles on a beach.'[4] In the centre of all this was the Emperor himself, splendid in his carriage of carved ivory with 'six jade-spangled horses, sleek as dragons'.[5] He watched the skill of his troops, the accuracy of their marksmanship, the tactics of his commanders; and after it was all over, he inspected the piled-up bodies of the slain, awarding praise and gifts to those who had distinguished themselves.

Such displays of military force kept the tribute states in their place. But the Han dynasty hunting parks were also used for magical ceremonies which kept supernatural powers equally submissive. At the end of the year the Emperor gathered his court to take part in warding off pestilences, when, according to Zhang Heng (78–139), a great crowd of boys and girls came out in black clothes and red turbans. With peach-wood bows they shot arrows into the sky, 'aimlessly, like a rain of stars,' and as the arrows fell they severed the demons' bridge to earth, beheaded the wheel-headed devils and brained the marsh goblins. When all was over the officer in charge went round the capital searching out any demons that had been left alive, until every house was so spiritually cleansed that 'nothing was left contrary to right and to rule'.[6]

The Park as Empire

The hunting park thus came to be a stage for ritual encounters between man and the cosmos, and so acquired added spiritual and magical associations which, somewhat tenuously, Chinese gardens would always retain. However, the parks were not merely settings for important ceremonies, but themselves began to take on a symbolic role. Again it was one which would become an important aspect of these gardens right up to modern times, for Chinese emperors came quite consciously to use the size and luxuriance of their parks to reflect the splendour of their dynasties, and rich men, following the imperial pattern, built lavish parks which gave form and elegance to their wealth.[7]

One of the first Emperors to make use of his hunting park in this way was Qin Shi Huangdi (the First Emperor of the Qin Dynasty) who, in 221 BC, conquered the last of the ancient kingdoms and joined them all for the first time into a united Chinese empire, with himself at the head. A ruthless conqueror, he sent the vanquished rulers and their families up to his capital and destroyed the palaces they left behind. Then he rebuilt them round his own palace as trophies of his victories. Beyond the city limits, too, he walled off a vast hunting preserve, the

'Magical palaces of the Immortals in the Mountains', silk tapestry. The Taoist book, Liezi, describes the Immortals' island mountains in the Eastern Sea covered in towers and terraces of gold and jade, surrounded by trees of pearl and garnet. Note the Immortals' storks flying among the clouds and the dramatic stones and boulders in the foreground.

Shanglin Park, in which he collected tribute of rare beasts and birds and trees from the vassal states. With this, the idea of the park as a microcosm of the empire began to be added to its earlier role as hunting preserve. In fact, when Qin Shi Huangdi's dynasty fell in 207 BC, and the empire was torn with wars and uprisings, the Shanglin Park was not destroyed. Already accepted as a symbol of imperial power, it was taken over by the new dynasty and is described in the great prose-poems of Han literature.

To praise the Emperor the writers of these works stocked his domain with creatures, both real and fantastic, brought from all corners of the known world:

> Unicorns from Jiu Zhen,
> Horses from Ferghana
> Rhinoceros from Huang Zhi,
> Birds from Tiao Zhi.

More and more the Shanglin Park seems to have become a magical diagram, a symbol of the Empire in miniature – although it was hardly miniature by our standards. Poets said that it covered a thousand *li* of hills and forests. (A *li*, or Chinese mile, is equal to about a third of an English mile.) Within it eight rivers converged symbolically from the four corners of the earth. From north, south, east and west they wound through forests of cinnamon and across broad plains, and swirled in wild confusion through the mouths of narrow gorges. Their waters were 'loud with fish and turtles'. Pearls and semi-precious stones lay glistening in the waves. Great flocks of birds flew wheeling and drifting in the spray.

In a series of masterly contrasts, Sima Xiangru (179–117 BC), court poet to the Han Emperor Wudi (the Martial Emperor), closes in upon the myriad flowers crowding the river banks, then backs away to show the mighty peaks of mountains rising beyond them. Now we see the whole powerful length of the rivers, winding through a landscape that is indistinct and magical, and now the minute scales of fish that lie motionless in their watery depths. In this description, where myth and fact are mixed in a way that suggests the teeming plenitude of the universe, we have the first flowering of the Chinese garden as a catalogue of earthly delights:

> The sun rises from the eastern ponds
> And sets among the slopes of the west;
> In the southern part of the park,
> Where grasses grow in the dead of winter . . .
> Live zebras, yaks, tapirs and black oxen,
> Water buffalo, elk and antelope . . .
> Aurochs, elephant and rhinoceros.
> In the north, where in the midst of summer
> The ground is cracked and blotched with ice
> Roam unicorns and boars,
> Wild asses and camels,
> Onagers and mares . . .[8]

Although in reality the imperial parks were somewhat less inclusive than this description suggests, they nevertheless enclosed extremely large tracts of land and a great variety of scenery. From Han times they also contained great orchards, farms and lakes which produced food for the court. To complete the total picture, the poets included descriptions of mountains, which, distant and awe-inspiring, their sides cracked with caverns and their peaks swathed in cloud, were clearly visible to the Emperor and his court from high stone terraces that rose above the trees and hills below. These mountains were essential if the parks were to be true representations of the Empire for in China great mountains were not just physically impressive, but thought to be the 'bony structure of the earth'. Even more potently, they were also centres of cosmic energy, the conductors of that magic electricity which could be seen flashing round their peaks, while the thunder roared and grumbled in the crags. From very ancient times such mountains were regarded with respect and even terror, while in the Tang dynasty (AD 618–907) as we shall see, the appreciation of rocks, themselves symbols of mountains, was to reach an almost spiritual intensity. Finally, miniature representations of mountains would find an essential place on every scholar's desk and in every pleasure garden. So striking, indeed, is the use made of rocks by the Chinese that it would be foolish to discuss their gardens without first looking at some of the myths and yearnings that have, over so many centuries, given mountains their peculiar fascination.

Dreaming of Immortality in the Mountains, attributed to Tang Ying, early sixteenth century. A Taoist hermit smiles in his mountain hut as he dreams of his immortal body travelling on the wind.

Mountains and the Legends of Immortality

Back in the fifth century BC, the author of many of the *Chu Ci* poems described a magical journey he made into the fabulous Kunlun mountains of the west. His strange and passionate poems reveal the mountains as intensely potent forms. As he glimpses the snaky branches of cassia trees entwined in a gorge he is appalled: 'How sheer and awful are the mountains! Shrinkingly my soul twists and turns its way through them'. And yet, though he crouches in caverns, paralysed with fear, the peaks hold him in a kind of trembling fascination. He soars above the clouds, half swooning with terror and delight, as if in the grip of a sexual passion:

> I mounted a high cliff's rocky walls,
> And stood at the woman rainbow's highest point;
> Resting on the sky, speeding on the rainbow,
> On I rushed till I touched the heavens.
> I sipped cool drops of refreshing dew,
> And rinsed my mouth with fine flakes of hoar frost.
> I went in the cave of the winds to rest myself,
> Woke with a start, gasping for breath.[9]

It is hardly surprising if the spectacular forms of China's mountains filled those who saw them with this kind of awe. Chinese painters do not exaggerate when they show cliffs rising directly from the plains, and summits floating half invisible above seemingly horizonless lakes. Even today, travellers in China seldom miss the intoxicating experience of seeing art confirmed by reality. The train from Luoyang to Zhengzhou, for instance, rattles along below the massive, sacred mountain of Hua Shan. To the left the plain stretches across a vast horizon, while to the right the land slopes steeply upwards. Far above – far higher than at first one thinks of looking – the jagged, violet peaks stand out against the sky, crisp as icebergs, their outlines soaring and leaping like the flight path of a lark. A million pilgrims must have toiled up these precipitous heights, with the mist drifting soundlessly below. Mountains such as this – one of the five that were thought to stand at each corner and in the centre of the Chinese world – received the homage of Emperors whose 'mandate of Heaven' was confirmed, in each reign, by a pilgrimage to the great Mount Tai

(Tai Shan). And because they might be inhabited by the famous Immortals of Chinese legend, mountains have also figured in the dreams of those who aimed at immortality.

The Immortals, or *xian*, were enchanted beings, yet not quite gods, who could dissolve in the air and fly about on the backs of storks. Their palaces were perched among the peaks of the mythical Mount Kunlun (the Himalayas) in the Far West, and on equally famous islands in the eastern sea, which forever melted away in the mist as human travellers approached them. Ultimately it is the idea of these magical dwellings that lies behind the rockpiles and strange standing stones of Chinese gardens.

Caverns in gardens, so useful in the summer heat, also had associations with immortality. The *xian* also inhabited another enchanted world, with its own sky, which existed in vast grottoes under the earth, grottoes which might even form a subterraneous link between all the five great mountains of antiquity. A few privileged human beings had descended to the palaces of this nether world:

> The entrance is so narrow, only a canoe can pass through. It gradually widens. . . . With blazing torches one goes down the stream, hearing ten thousand flutes blown by the wind. . . . After one hundred *li*, a hole for fish. Then a universe lit by sun and moon. Peaceful grasses, harmonious clouds. Birds answer the call of men, flowers listen to visitors. It is a complete other world.[10]

These enchanted habitations were the particular concern of the Taoists, or to put it more accurately, those seekers after immortality who had adopted the name of Taoist around the first century AD. The desire to be immortal had ancient roots in China, an optimistic view of life very far from the Christian or Buddhist idea of this world as a vale of tears. For the Chinese immortality-seekers wished to achieve physical immortality, deeming it safer, no doubt, to hang on to the body than to trust to the less tangible future of a soul. Their aim was not absolute immortality, but a more modest extension of life: about two to five hundred years.

To achieve this these Taoists plunged into a busy programme of activities designed to refine the body – activities that embarrassed some later historians of philosophy who found magic a

little hard to take. The programme included not only abstinence from eating grains and Yogic breathing exercises, but also highly complicated sexual techniques involving numerous (and if possible virgin) partners, and the collection, preparation and ingestion of various herbs. Mercury was also an important ingredient, so that the elixir of life quite often proved deadly: at least two Emperors of the Han dynasty died from over-enthusiastic doses of immortalizing drinks. Nevertheless, over the centuries, the Taoists gathered an extraordinary range of empirical scientific knowledge as they scoured the world of plants and minerals for their potions. And their search also took them up into the mountains where they wandered alone, gathering herbs and the life-prolonging *lingzhi* mushroom.

In Chinese art the Taoist adept can usually be recognized by the sacred fungi represented somewhere near him. Additional identification may be provided by two symbols he often carries. One of these is a narrow-mouthed, double-bodied gourd, an emblem of the Other World of the Immortals which is often incorporated into gardens in the designs of pebble paths and gateways. The other is an image of the sacred mountains hanging from his staff. Such miniatures played an important role in the Taoist's bid for immortality. By recreating a mountain on a reduced scale, he could focus on its magical properties and gain access to them. The further the reproduction was in size from the original, the more magically potent it was likely to be – a process not unlike that of a cook who boils down his stock to concentrate its juicy essences. Representations of potent sites in miniature were thus not aesthetic in origin, but were pieces of practical magic.

Today connoisseurs in Japan and Hong Kong still gain much satisfaction from tiny landscapes created in shallow pottery and porcelain bowls. In mainland China too, the old, related skill of dwarfing trees continues. Although the magical Taoist meanings of this practice have been submerged by centuries of skilful practice and artistic criticism, one cannot help feeling that an atavistic sense of its magical origins persists, however deeply sunk in the Chinese subconscious. Dwarfing a living tree was accomplished by slowing down the sap, just as the Taoist's respiratory exercises slowed down the journey of breath around his body. For both, the technique was to twist and stretch the limbs. The little trees grew into strange contorted shapes, bent like the

ABOVE, LEFT *Miniature landscape in a bowl, here on show in the Canton Railway Station. The viewer shrinks himself in imagination and wanders undisturbed by the discomforts of real mountain travel. Purists disapprove of miniature huts or figures in these 'gardens' since they are too literal, and thus limit the imagination.*

ABOVE, RIGHT *A tree in a Shanghai nursery garden. The trunk has been bent and wired until it has grown into this knot, representing the character* shou *(longevity).*

OPPOSITE
ABOVE, LEFT *A dwarf tree displayed in a courtyard of the Liu Yuan, Suzhou.*

ABOVE RIGHT *Flower stand in a Hangzhou nursery garden. Tables and chairs are also made from these contorted roots, and are often placed in garden pavilions where they suggest the fairyland of the Chinese Immortals.*

BELOW, LEFT *Tree trunks in the Yu Hua Yuan (back garden of the Imperial City), Peking. Distortion and twisting suggest age, and thus immortality.*

BELOW, RIGHT *Dwarf trees in the Longhua Botanic Gardens, Shanghai.*

bodies of those who wished to enter the cavernous 'Other World', and looking as aged as were the Immortals themselves. In large gardens certain areas are still set aside for growing such miniatures, while famous centres of garden art, such as Suzhou, continue to develop their own particular styles of distortion under families of master craftsmen.

Mountains Enter the Garden

It was to represent the homes of the Immortals that mountains first became a theme for objects of art in the incense burners of the Qin and Han dynasties. Made of pottery, or bronze inlaid with gold, their layered peaks concealed small holes. When incense was burnt inside them, it seeped out among the tiny crags like scented mist. Shrinking himself in the mind's eye, a man could then imagine that he was among those mysterious islands of the eastern sea, often sighted by sailors but never yet visited by man.

In the third century BC, China's 'Great Unifier', Qin Shi Huangdi, had already sent an expedition of young men and maidens to find this archipelago and bring back the secret of eternal life. For despite, or perhaps because of, his immense worldly success, he did not take kindly to the prospect of death. At his court shamanistic magic flourished, and the cult of immortality first appeared as a clear historical phenomenon.

As it happened the great conqueror died of natural causes, ironically while travelling to the eastern regions in pursuit of the elixir of immortality, but the fruitlessness of his quest did not deter his successors from pursuing the same dream. One later Emperor who was particularly taken with the subject was Han Wudi (141–86 BC), and it was under his auspices that representations of the Immortals' mountains first appeared in the imperial parks. Wudi was convinced that the Islands of the Immortals were not a myth. If Qin Shi Huangdi's expedition had failed in its purpose, that must be because the *xian* did not care to have human visitors trampling about on their movable paradise. Instead of going to them, Wudi therefore decided to entice them to come to him. In the grounds near the palace he would build replicas of the Immortals' abodes, in which the three islands, Penglai, Yingzhou and Fangzhang, would be recreated in so ravishing a form that the Immortals would mistake them for their true homes. Once there, in the palace gardens, they could perhaps be persuaded to give up the secrets of longevity.

Thus, in the Qianzhang Palace, Han Wudi constructed two spacious lakes with islands. Here were built simulacra of the fairy mountains, arranged 'in a pattern'. Beyond them arose distant views of the hunting parks. The isles themselves were planted with remarkable flowers that bloomed even in winter, 'spirit trees' in groves and jagged rocks that looked like crags. 'And all around were herbs of mystic potency, and in the lakes, huge fish, which sometimes got caught in the shallows.'[11]

A barbarian shaman advised on the necessary techniques for creating suitably potent buildings. The devices he recommended included a revolving weather vane and a maze of corridors and pavilions with no fewer than a hundred doors and a thousand windows. On high columns, statues of the Immortals held up bowls to catch the dew, 'reaching out beyond the defilement of [this world's] clogging dust to gain the limpid elixir of pure, transparent *qi* (i.e. dew, here identified with the vital breath, essential element of all things).[12] In fact, as the historian-poet Ban Gu (AD 32–92) remarked somewhat disapprovingly, the Emperor was carried away by the magician's 'windy verbiage', and created 'an abode for immortals – not a place where mortals can find peace'.[13] Nevertheless, across the centuries Han Wudi's fairy islands have spawned countless miniature mountains built in gardens all across the East. In Japan they led eventually to the symbolic rocky isles of garden lakes, and in all Chinese gardens from this time on we will find echoes of the old Emperor's ambitious dream.

The First Private Pleasure Gardens

While the rulers of China were thus engaged in tempting the Immortals, many of their more successful subjects were also beginning to make themselves pleasure gardens. From the start these gardens showed two distinct and contrary tendencies: one towards the simple 'scholar's retreat', the other towards a more ostentatious and elaborate display of wealth. On the whole the simple garden style seems to have developed more straightforwardly from those plantations of useful fruit trees and hardwoods which we have already seen growing near the houses of the *Shi Jing* poems. Elaborate private gardens, on the other hand, took their style from the imperial parks. On the whole too – although this is a somewhat free interpretation of the evidence – the elaborate style was favoured by princes and aristocrats, and by *nouveau riche* merchants who prospered under the Han, while the simple garden appealed more to the new class of government administrators that began to develop from the beginning of the dynasty. In time it was the descendants of these administrators who developed the Chinese private garden with all its wealth of meaning, so it is perhaps worth a small detour to see how, as a class, they first began. The founder of the Han dynasty, Han Gaozu (206–195 BC), was a commoner. When he came to power he wanted officials who could curb the power of his own military commanders, yet who were unconnected to any previous ruling group. He found them among non-noble landowning families, who now came streaming into the capital to take up government posts. Down in the country they left one branch of the family on their hereditary estates, so setting in motion a relationship between city and country that, in China as in England, has always been regarded as

vitally important to the moral strength of the elite. Once in the city, these new administrators looked for a philosophy of government: they rejected the harsh legal policies of the previous dynasty and turned instead to Confucianism. Under the Han this developed into a powerful tool of imperial government and became the ruling orthodoxy of the state and its administrators for the next two thousand years.

We know these men had gardens from a chance remark about the great Han scholar Dong Zhongshu (179–104 BC) who, it was said, worked so hard on the Confucian Classic *Spring And Autumn Annals* that he had no time to visit his garden for three years. That he had a garden at all causes no comment, only his strength of purpose in taking no time off to visit it. One can conclude, then, that gardens were common enough among Confucian gentlemen, who usually found time to enjoy them. It also sounds as if the scholar's garden was some way away from his house or place of work, perhaps beyond the confines of the city, but unfortunately no further details of these gardens exist. Nevertheless it is possible to speculate a little on the qualities a good Confucian might have looked for in a garden.

The Confucian Connection

Although Taoism is usually thought of as the philosophy in China most connected with man in nature, Confucius also in his own way created links between the simple life of farmers and proper daily conduct, so that he too was instrumental in forming attitudes which find some part in garden design.

Confucius believed strongly in agriculture as the basis of the state, and looked back to the golden age of the mythical farming Emperors Yao and Shun, when refinement and simplicity were in perfect balance. Hence Confucianism, while elevating ritual, scholarship and 'the superior man', also celebrated the virtues of the 'simple and ancient' which were expressed in a single word.[14] The most important ritual performed by the Emperor each year was ploughing a furrow to symbolize his involvement with the soil. The prosperity of the state was thought to be dependent on his personal virtue, and his personal virtue included the frugal simplicity of farmers as well as the high refinement of a gentleman.

This Confucian ideal found expression in gardens side by side with more mystical notions of immortality. Behind a magic mountain it is not unusual to find a country cottage surrounded by an orchard of apricots. For instance in the Summer Palace near Peking there is still a row of farm buildings on Longevity Hill. Today it is somewhat dilapidated, but, in the time of the Qianlong

RIGHT *Pavilion of Joy in Agriculture; a simple farmhouse set among the trees on the eastern slopes of the Wan Shou Shan (Longevity Hill), Yi He Yuan (Summer Palace), near Peking.*

BELOW *'Composing Poetry on a Spring Outing', attributed to Ma Yuan (active c. 1190–1225). Although the landscape seems 'natural', the presence of a table under the trees, and the elegant pavilion half-visible on the right, make it likely that this 'outing' takes place within the garden of one of the company, rather than far away in the hills.*

Emperor it was a little working farm. The old Empress Dowager Cixi also used to enjoy watching her ladies struggling to manage chickens on her farm, while an important annual festival for Empresses involved the ritual tending of mulberries for silk worms.

In some of the most elaborate Suzhou gardens there are also allusions to Confucian simplicity in the names given to pavilions and terraces. In the Zhuo Zheng Yuan one summer house on an island in the lake is called 'Persuading One to Farm Diligently', while a later scholar even justified building gardens – quite seriously it seems – on the grounds that a farmer passing by a spray of spring blossoms would thereby be reminded it was time to start the sowing.

Gardens of the Nouveaux Riches

If reasonably simple gardens were acceptable among the Confucian gentry during the Han dynasty, elaborate gardens were still rare enough to cause comment. A certain merchant, Yuan

Guanghan, for instance, owes the perpetuation of his name entirely to the fantastic garden he made on the outskirts of the capital. He was the proud owner of eight hundred slaves and a beautiful piece of land stretched out at the foot of a mountain famous for its excellent stones. Here he had streams diverted through a park extending four *li* from east to west and five from north to south. Within the park were forty-three fine halls and terraces linked by open galleries – a squeeze, even for such generous perimeters. Like the imperial parks, Yuan Guanghan's was stocked with rare beasts and trees. Parrots flew overhead. Oxen roamed around freely. Tibetan yaks strolled by. But by far the most interesting feature was an artificial mountain said to be a hundred feet high, the equivalent today of a ten-storey building.[15]

The fame of Yuan Guanghan's garden has left us the first record of an artificial mountain, but for him it was a disaster. Perhaps his extravagance amounted to *lèse-majesté*, rather as Fouquet's palace at Vaux-le-Vicomte challenged the supremacy of the Sun King. Or perhaps the Emperor feared the Immortals might descend into the wrong garden. At any rate Yuan Guanghan, like Fouquet, soon found his life was under investigation. His garden was appropriated, his possessions were confiscated, and finally even his head was removed.

Despite this example, however, there has been no lack of extravagant gardens in Chinese history. Countless men have followed in Yuan Guanghan's footsteps and fabricated parks which ostentatiously displayed their wealth – although not all of them ended by losing everything. In Wuxi today there is a very recent garden, begun in the 1920s, which is a prime example of the tradition. It is now a public park, with some graceful pavilions added to it in the 1950s. But its most noticeable feature is a gallery that winds along the shore of the lake on which it is sited, and ends up at a miniature concrete pagoda. The other end of the gallery leads past a rock mountain covering at least half an acre, and perhaps twenty feet high at the top. For me this pile has all the signs of ostentation; grotesquely eroded stones stuck together with concrete, unnecessarily heavy massing, over-obvious likenesses to animals or distorted human beings. It is a garden that speaks more of its maker's wealth than sensibility, and it is a long way from the grace and delicacy of a garden such as the Wang Shi Yuan which we looked at in the first chapter. Perhaps the difference lies quite simply in the aim of the gardener: the one to display wealth and receive exclamations of amazement; the other to express his own delight in the forms available to him, to create a place he himself found beautiful.

For many garden makers the motives were perhaps a little mixed. The creator of the simplest garden is unlikely to object to praise from those he admires, and as we shall see, Chinese gardens were above all places for social and literary gatherings. Much of their point, then, was to give enjoyment to many people. This was true of both the largest imperial estate and the smallest and simplest private plot.

The next development of the Chinese garden was one that deeply affected all the arts in China, particularly poetry and painting. It was bound up with the long, drawn out collapse of the Han dynasty and a corresponding loss of faith in Confucian values. It first affected private gardens, and only later, through them, came to influence the parks of the Emperors. Since, however, we have left imperial gardens somewhat hanging in the air, and since private gardens are about to develop in a new direction, it seems best to treat them in two separate chapters to avoid too great a confusion. The next chapter then will tell the story of the great imperial gardens up to the present day. The following chapter, covering much of the same time span, will go back to the end of the Han dynasty and follow the development of private gardens, both simple and grand, as a separate tradition.

Rockery, Li Yuan, Wuxi. A labyrinth of rocks which rise into an enormous pile topped by a concrete pavilion. Several alternative paths lead up to it through an underground cavern, tunnels and deep rocky chasms. Along these routes particularly strange rocks have been cemented above the others, and stand out like rows of contorted animals. The whole effect is highly artificial and to Western eyes not attractive. But the view from the top of the 'mountain', with garden and lake stretching out to the horizon, is beautiful.

IMPERIAL GARDENS

On the morning of 18 October 1860, a detachment of British troops marched briskly out of one of the city gates of Peking. Heading northwards along the paved road made for the Kangxi Emperor a hundred years before, they came in time to two fan-shaped lakes, so designed that in the height of summer they would suggest cool breezes wafting towards the entrance of the imperial gardens of the Yuan Ming Yuan. Ten days earlier this detachment had left these same gates as part of a much larger force of combined French and British troops. At that time they had been laden with booty – all the movable remains of a century of imperial collecting, and behind them they had left the palaces looted, their treasures smashed and scattered in the open courtyards. Now they re-entered the gardens without the need of force and, under orders from their officers, spread out among the sixty thousand acres of water gardens enclosed within the walls. Then they set fire to everything that was left. In two days of burning they thus destroyed almost two thirds of all the three thousand varied buildings that had once made this place the talk of Europe and one of the glories of the Manchu dynasty.

In England only sinologists and art historians are now likely to have heard of the Yuan Ming Yuan. But in Peking it is still remembered, for, just to the north of the city down narrow bicycle paths, the remains of one small part of it can still be seen among neat vegetable fields and groves of newly planted trees. And from the top of little hills there, which once were islands in the waters of the Happy Sea, the outlines of its streams and valleys are dimly visible beneath the local farmers' lotus beds.

There also still exists the Yi He Yuan, rebuilt after the fires and known to the West as the Summer Palace. In size it covers less than a quarter of the original parks, but from its courtyards and the terraces above its great artificial lake you can still get a flavour of the old gardens as they must have been. Even there, however, reminders of the destruction remain. On the northern slopes of the Wan Shou Shan ('Longevity Hill'), there are still a number of roofless temples and crumbling, dusty red walls. These, the guides may tell you, are remains left by the British – a supposedly civilized country that, as a Chinese writer once remarked, 'makes war like a barbarian.'[1]

A View from the Other Side

For the British Commander Lord Elgin, however, the sacking of the Yuan Ming Yuan was a necessary evil, inspired, indeed, by humanitarian feelings. The Chinese had inflicted appalling tortures on British prisoners of war, and whatever the rights and wrongs of the conflict, Elgin fervently believed that such horrors could not be allowed to pass unchecked. By burning the gardens, he felt he could hit directly at the Emperor himself, the man ultimately responsible for the conduct of his troops, without causing unnecessary misery to his innocent civilian subjects. But Lord Elgin knew little of China. To the inhabitants of Peking it was unthinkable that a small force of foreign troops should be in a position to chastise the Chinese Emperor and, moreover, they also had a far more reasonable and traditional explanation for the burning.

Chinese history is littered with the corpses of gardens. From very early times Chinese folk tales have connected the growth of imperial luxury with the inevitable decline of great dynasties, and as often as not the symbols of such luxury have been beautiful women and imperial parks. It is a connection confirmed by history. The end of dynasties has almost inevitably been accompanied by the shrieks of women, the crash of falling rockeries and the crackle of burning pavilions.

In this case a great peasant rebellion had recently ravaged the whole of south China while stories of the Xianfeng Emperor's latest dissipations circulated in the market places. Thus, when on that October morning a great pall of smoke rose up over the towers of the Yuan Ming Yuan and began to drift slowly eastwards, dropping splinters of burning wood, hot ash and cinders into the courtyards and alleyways of Peking, it was in many ways a well-established confirmation that the Manchu dynasty was nearing its end.

Ancient Dissipations

It is just possible that the origins of this association of luxurious gardens with imperial dissolution go back to the end of the earliest Chinese dynasty – the shadowy and perhaps mythic Xia – which is supposed to have come to an unpleasant end around 2500 BC. Later, whenever conscientious ministers found their monarchs planning gardens, they seldom failed to remind them of King Jie, whose sadistic dissipations were said to have cost him the throne. One classic story about him is that he set up immense troughs of rice wine, at great cost to the people, and his drunken courtiers tumbled into them as they bent to drink. In some versions these troughs became a whole artificial pond of wine on which King Jie went boating, poled by suitably lascivious courtesans.

The next dynasty, the Shang, also succumbed to luxury, and fell in its turn around 1050 BC. This time there is no doubt about the part played by gardens. Looking back later, the philosopher Mencius made a direct connection between the rulers' lavish parks and their moral decline, a connection later Emperors might have done well to remember. 'They pulled down houses in order to make ponds,' he said, 'and the people had nowhere to rest. They turned fields into parks, depriving people of their livelihood.'[2] This loss of agricultural land precipitated moral, then social, and finally even physical collapse. 'With the multiplication of parks, ponds and lakes, arrived birds and beasts. By the time of the great tyrant Zhou the Empire was again in disorder.' This tyrant, a kind of Chinese Nero, spent his days collecting horses and dogs and rare objects, held orgies all night, refused all advice, and, most significant of all, 'extended his parks and terraces without limit'. Once again they were a symbol of his excesses, which eventually led to the fall of the dynasty.

Interestingly enough, Mencius does not condemn landscaping out of hand. In an earlier passage he contrasts the great park made by the wise and magnificent King Wen of Zhou with the smaller but much more socially disruptive grounds of King Xuan of Qi. Though the former covered no less than seventy *li* and contained a manmade terrace and lake, these seem to have been open to the people and were used productively: 'When the Ling Terrace was built,' said a later Emperor, 'the common people came to enjoy it. When the Ling Pond was made, there were plentiful fish.'[3] The word *ling* means 'mana' and refers to a beneficent power which, in China, was often concentrated in powerful-looking rocks and other natural but unusual phenomena. Here it suggests that apart from their aesthetic and economic value to the people, the pond and terrace held some magical properties. Reflecting a ruler's desire to place his resources – both secular and supernatural – at the service of his subjects, King Wen's park was regarded by Mencius as a proper expression of imperial rule.

Marble terrace, Yu Hua Yuan (back garden of the Imperial City), Peking. A faint memory of the terraces of ancient kings is still suggested by this marbled perch among the trees. A stable is concealed in the rocks below with, on the far side, a small outdoor run so the animals kept there could be seen by garden visitors. Menageries were common in grand gardens, again a vestigial reminder of their origins in imperial hunting parks.

The reign of King Wen was an exception, however. More typical was the much later story of the Sui dynasty which, in AD 581, re-united China under one rule after years of disorder. In this case all Mencius' worst fears about excessively lavish pleasure grounds were well justified. The founder of the dynasty was a tough, energetic and parsimonious official who concentrated all his energies on getting the Empire together again. His heir, however, has found fame as a classic example of the dissolute last ruler. Known to history as Sui Yangdi, he is said to have found in the idea of imperial greatness unbounded possibilities for the expression of megalomania. He sent three great military expeditions to conquer Korea, and ordered the building of a grand canal linking the rice country of the south with the northern capitals. As soon as he ascended the throne he started building Luoyang, a second eastern capital to supplement his father's old capital at Chang'an. And here, to signify the might and glory of his house, to equal and even surpass the parks of Han, he enclosed a landscape park seventy-five miles in circumference

In this he ordered a lake to be dug almost six miles long. From its glassy expanse three Isles of the Immortals, lavishly endowed with pavilions, rose in imitation of Han Wudi's, and waterways connected the lake with further pools and streams to symbolize the five lakes and the four seas of the ordered universe. The shoreline wound through a landscape with 'a thousand prospects and a variegated beauty unequalled in the world of men'.[4] To stock the hills and valleys of this huge domain, an imperial proclamation commandeered any unusual plants from neighbouring estates, including fully-grown trees, which were transported to the park on specially constructed carts. Within a few years 'the garden's verdure, trees, birds and animals had multiplied to the point of luxuriant abundance. Peach walks and plum byways met beneath kingfisher-green shadows. Golden gibbons and green deer sped by . . .'

So far the account suggests something like a re-creation of Han splendour, magnificently celebrating the glorious restoration of imperial unity. However, the park soon became a synonym for outlandish extravagance. As a young man Sui Yangdi had been made Viceroy in south China, and here, according to northerners, he had developed a taste for luxurious refinements. In his new pleasure grounds, instead of imitating the old Han-style hunting lodges, he now built sixteen water palaces strung out like a necklace of pearls along the lakes and waterways. Each was surrounded by a garden of its own more elaborately and artfully decorated than the surrounding park. No fantasy was ignored to make these smaller gardens perfect at all times: 'In autumn, when the leaves had fallen from the maples, the trees and bushes were decked with leaves and flowers made of glistening fabrics. In addition to real lotuses, the lake was adorned with artificial lotus blossoms which were continually renewed'.[5] The palaces could be approached only by water, on imperial barges with prows that curled up to form dragon heads. Each palace contained twenty picked concubines, skilled in the arts of singing, playing music, versifying and dancing, and also, no doubt, in entertainments of a more fleshly nature. Diversion was also provided by a remarkable series of automata. Guests were seated along the edge of specially constructed channels while mechanical figures, two feet high and sumptuously dressed, passed by in front of them in boats. Some of the automata were designed as singing girls, while others performed no less than seventy-two different mechanical scenes from Chinese myth and history.

On good authority it is said that 'a million' people worked to create this park. One scholar writes that five out of ten died in its feverish construction and that the same proportion was lost during the building of the great canal and the disastrous expeditions against Korea.[6] In AD 616 the inevitable rebellion broke out among the populace. The Emperor retreated to the city (now called Yangzhou) at the confluence of the great canal and the Yangtze river, leaving his generals to fight over the Empire. In 618 he was assassinated by his own bodyguard, who first stabbed to death his favourite son.

The archetypal nature of Sui Yangdi's story makes it read like a morality tale; indeed, the thirty-seven years of Sui rule partly encapsulate the histories of all dynasties. It is little wonder then, that the founder of the new Tang dynasty (618–907), took an immediate and firm line on landscaping. He even led his officers and relations on a cautionary expedition to the wrecked palaces of his predecessor: 'I do not want you to dig ponds! Nor to make gardens! Nor to build pleasure parks at the expense of farmers! I forbid you to indulge yourselves!'[7] His words blew away on the wind. He himself found it expedient to build a 'Palace of Great Brilliance' suitable to his imperial ambitions, and included around it extensive grounds for royal entertainments. Within a few generations his successors and their advisors were all back at the old game, vying with each other in the lavishness of their gardens.

More Ways to Elaborate

The morality tale of gardens and collapsing dynasties continued in China right up to the burning of the Yuan Ming Yuan, more than twelve hundred years after the fall of Sui Yangdi. Although they varied in size and splendour, all the subsequent imperial pleasances were nevertheless on a very large scale, and right to the end continued to shelter exotic plants and beasts symbolic of the Empire's riches. Over the centuries, however, other themes also influenced their development, and none more fundamentally than the ideal of the 'scholar's retreat'. This, as we will see in the next chapter, developed particularly after the fall of the Han dynasty, and was in many ways a reaction against extravagant gardens. In time, however, the ideal of the garden as a simple but refined rural retreat found expression even in the most luxurious imperial parks.

At the end of the seventh century, for instance, the notorious and extraordinary Empress Wu Zetian (the only woman incidentally who ever held the full ruling title of Empress) decided she liked the simple life. She took a fancy to a palace in the forested mountains of Shaanxi, some sixty miles north of her capital, where she moved in summer with the ministers of the realm trailing uncomfortably after her. The park occupied much arable land. It was unfenced, a lurking place for robbers, and difficult of access. Moreover the accommodation was so inadequate that half the court had to sleep in grass huts. From here the business of the realm was expected to function as smoothly as it did from the capital. In 700 the ministers complained: the situation was impossible. But rather than forgo her country palace the Empress took this as a golden opportunity to rebuild the whole estate at vast expense. There was little the ministers could do then but thank the lady who thus showed so much concern for their welfare.

There were few men brave enough to protest against imperial extravagance. Joseph Needham tells a story of the censor Zhen Zhijie, who in 747 sent up a powerful admonishment to the Emperor Ming Huang of Tang on the subject of his water gardens. For some time the minister received no reply, which made him increasingly apprehensive. Then, on one of the hottest days in summer – and in central north China that means heat like a furnace – he was summoned to appear at midday in a pavilion among the very gardens he had criticized. It gives a remarkable picture of how elaborate were the effects attempted at this time:

> The Emperor was in the Cool Hall, and behind his seat the water struck the fan wheels while cool air played around his neck and clothes. Zhen Zhijie arrived and was given a seat on a stone chair. A low thunder growled. The sun was hidden from sight. Water rose in the four corners and, forming screens, fell again with a splash. The seats were cooled with ice, and Zhen was served with marrow-chilling drinks so that he began to shiver and his belly was filled with rumblings. Again and again he begged permission to leave, though the Emperor never stopped perspiring, and at last Zhen could hardly get as far as the gate before stooping to relieve nature in the most embarrassing way.[8]

Gardens of Elegance, Music and Love

This Emperor, Ming Huang ('the Brilliant Emperor'), had ascended the throne in 712 when he was twenty-eight years old. He was vital, intelligent and decisive, with a strong vision of his role and of the greatness of the Empire. Moreover, it was a vision that included patronage of the arts, as well as territorial greatness. In the Pear Garden conservatory the Emperor himself instructed the court musicians, and at his court poetry flourished as never before or since in China. Indeed, for the first fifteen years of his reign this spectacular Emperor maintained a balance between artistic patronage and personal luxury that made his court a model of elegance and grace. Then, so the story tellers like to say, he fell in love. The lady, an exquisite imperial concubine named Yang Guifei, is one of the four most famous beauties in Chinese history, since for her Ming Huang threw away his Empire.

As he grew older Ming Huang grew increasingly enamoured, not only neglecting the cares of state for his mistress' company, but giving in to her every whim. Her innumerable relations

soon occupied all the positions of power, and the Emperor spent his days building palaces for his lady and her sisters. Each had its own elaborate garden quite different in style and detail from the more natural imperial parks beyond the city walls.

Hua Qing hot spring, near Xi'an, (ancient Chang'an), Shaanxi. Although the buildings are modern, people still come to bathe here as they did in the time of Yang Guifei.

One of the most famous of these gardens still exists, although nothing remains in it from the time of Yang Guifei. It is situated some twenty miles east of modern Xi'an, on the site of the ancient gardens of the Hua Qing hot spring – already famous for a thousand years before Ming Huang first went there. Here, according to a modern scholar, Ming Huang 'built a microscopic island mountain of lapis lazuli, around which the girls of his seraglio sculled boats of sandalwood and lacquer. This rich and splendid piece of landscaping represented the height of aristocratic fashion in gardens'.[9] Here too the handmaidens of Yang Guifei are said to have bathed her in a marble pool while the besotted Emperor looked on through a secret peephole.

Despite his softening over the years, Ming Huang was a great monarch, and his reign (712–756) is regarded as one of the high points of Chinese civilization. But it is agreed by scholars that the most cultivated of all Chinese Emperors was probably the Emperor Huizong (1100–1126) of the following dynasty – a man of exquisite refinement, a justifiably famous painter and a garden maker to beat them all. The Empire paid a high price for him however, as it had for Ming Huang. Each of these Emperors abdicated and both died miserably, their personal lives shattered and their domains in ruins.

Birds and flowers attributed to the Emperor Huizong, Song dynasty (detail). Scent and taste are suggested in the choice of gardenia blossoms and lichee fruit: sound by the little birds that are perched delicately among them.

The Tang dynasty, however, was not totally destroyed by the catastrophe of Ming Huang's rule: after a break of only eight years it was restored and survived another hundred and fifty years after Ming Huang's demise. But under Huizong the Song dynasty permanently lost the whole of north China to the half-savage Jurchen barbarians. To an extent unparalleled before or since, this collapse was due to an imperial passion for landscaping. In fact it is one of the ironies of Chinese history that this *ne plus ultra* among cultivated monarchs can be said to have sacrificed his civilization for the sake of a garden.

Huizong and Petromania

Painting was undoubtedly Huizong's main talent. His delicate coloured studies of birds and flowers are among the loveliest in all Chinese art, and his ink paintings were said to have reached the class called 'divine'. His favourite subjects were quite specifically connected with gardens, or at least with details of nature to be seen in gardens – a plump white pigeon puffed out on a spray of plum blossom, a parakeet on a twig of peach. They are exquisite works, for Huizong was a supreme perfectionist. He used to take out members of the Imperial Painting Academy to observe their subjects first-hand in the palace gardens, and he was such an obsessive realist that he is said to have once rejected the painting of a pheasant because the artist had failed to notice how, when stepping on to a mound, this bird always raised its right and not its left foot first. The same obsessiveness characterized his planning of the imperial garden at Kaifeng; for if painting was his greatest talent, gardening was his abiding passion.

Like other imperial landscapists before him, Huizong had all the wealth of the nation at his disposal and, as a source of inspiration, the whole tradition of imperial parks with their Isles of the Immortals and their collections of rarities from all over the Empire. In addition, he had two other models which by this time were both highly sophisticated. One was the elaborate, highly decorative garden style of great officials and aristocrats, the other those simple naturalistic creations of poets and philosophers which had influenced the taste of Empress Wu. Both these quite different traditions of garden design had flourished in the previous (Tang) dynasty, and also in the early years of Huizong's own ruling house. All three now found some part in his new design. Thus, as in all imperial parks, there were trees and animals from across the Empire, including lichees from the south and golden pheasant and deer in the usual remarkable quantities. But also, less extravagantly, there was a useful herb garden and a rustic farm growing beans and grain. They were a reminder to those who saw them of the simple joys of agriculture – unmarred, of course, by any experience of toil.

By the standards of other imperial gardens Huizong's was not especially large. It did contain, however, as its most spectacular achievement, a feature which, though an accepted element in Chinese gardens since the Han dynasty, was here attempted on an unprecedented scale. Huizong was above all a lover of stones, a petromaniac of the first order; and what he actually built to the north-east of his capital, which stood on a great flat plain, was the gigantic recreation of a mountain landscape, an artificial pile more than ten *li* in circumference, of 'ten thousand layered peaks', with ranges, cliffs, deep gullies, escarpments and chasms. In some places the structure rose two hundred and twenty-five feet above the surrounding countryside,

'A literary gathering' (detail), attributed to the Emperor Huizong, Song dynasty. A garden party with tables set out in the imperial park. The setting is very similar to that in the background of 'Eight Riders in Spring' (page 110).

ABOVE, LEFT *Naturally formed mountain in Guilin is domesticated by rough steps half obliterated by moss and foliage. The Emperor Huizong may have wanted to make his false mountain look something like this.*

ABOVE, RIGHT *Deer-shaped tree, symbolizing long life, nursery garden, Hangzhou. Huizong's dragon-twisted trees may have looked something like this, although perhaps more spectacularly contorted.*

and in others it fell away, through foothills of excavated earth and rubble, to ponds and streams and thickly planted orchards of plum and apricot.

Four contemporary accounts of this astonishing place remain, one of them by the Emperor himself. These descriptions do not always quite tally with one another, but they make good reading. To the east, it seems, the Emperor could stand on a high ridge looking clear over the tops of a thousand plum trees, their scent in spring rising from below on the warm air. Bowered in all this fragrance were various buildings, among them the 'Hall of the Flower with the Green Calyx', a library and a circular pavilion for the Immortals. High on the hillside was a smooth and gleaming precipice of purple rock, which was reached by stone steps winding up along the cliffs. As the Emperor passed by, toiling minions opened a sluice gate in the hill and a manmade waterfall tumbled momentarily across the rock beside him. From this viewpoint in spring, the mountainous peaks beyond must have looked as if they were floating on the orchard blossoms. Their outlines rose in tiers, forking into two peaks, one to the east and one to the west, jointly known as Longevity Mountain.

However, even more magical than these general effects were the collections of incredibly shaped rocks that spiralled out of the hillsides at various points. For instance, along a ridge that connected Longevity Mountain with another pile to the south, boulders resembling animals and bizarre heads had been so placed that they looked as if they had been sculpted there by mountain streams: 'They were all in various strange shapes, like tusks, horns, mouths, noses, heads, tails and claws. They seemed to be angry and protesting against each other. Among them were planted gnarled trees and knobbed, clinging vines and evergreens.'[10] A little further on, this

same fascination with the grotesque gave rise to a display of fantastically contorted pine trees, 'their branches . . . twisted round and knotted to form all kinds of shapes, like canopies, cranes, dragons'.

The finest rocks of all were displayed along the imperial carriage road at the western entrance to the garden. One of these, fifty feet high, stood in the centre of the road with a little kiosk of rocks tied together to guard it. 'Other rocks on the side had various forms. Some looked like ministers having audience with the Emperor. They were solemn, serious, trembling and full of awe. Some were charging forward as if they had some important advice or argument to present.'[11] Huizong was so intoxicated with these remarkable anthropomorphic stones that he gave them names which were engraved onto them. The finest of all had their names written in gold, and round them clustered lesser rocks like courtiers.

From the highest point in the whole composition the Emperor beheld what seemed to be a microcosm of the universe. He looked down across the moat, wine shops and bamboo groves of the city suburbs, which spread out below 'as if lying on the palm of the hand'; and on all sides he could see 'peaks, caves, kiosks, pavilions, mature trees, and grasses, some tall, some short, some far, some near; some flourishing, some withering'.[12]

Beauty Confirms Virtue

Huizong's favourite pastime after official business was to wander in this garden. Here, without the discomforts of travel, he could find excitement and beauty enough to wipe out the memory

ABOVE *Lake Tai, looking over the Li Yuan, Wuxi. The most highly prized rocks came from this lake. They were modelled by the action of water grinding small hard pebbles into the larger, softer stone. In Ming times and later suitable pieces of rock were dropped into the lake to be 'harvested' by later generations.*

RIGHT, ABOVE *Monolith, Mei Yuan, Wuxi. The cost of transporting such vast boulders up to the imperial rock garden at Kaifeng helped to bring the Northern Song dynasty to an end.*

RIGHT, BELOW *Working on a new rockery, Wuxi, 1975.*

(or so he claimed) of all his pomp and splendour. Often, he would 'go up the hills or plunge into the valleys, exploring into the deep and dangerous', enjoying the *frisson* of terror he experienced while walking across deep chasms spanned only by a rustic log. It seemed to him that his artificial cordillera held a greater range of experiences than any real mountain in China. It blended with nature 'as if it had been here since Creation', but above all it was also a paradise, like the Land of the Immortals. One day the mushrooms of immortality were found growing on one of the peaks, a sure sign to Huizong that the forces of heaven were on his side. At any moment the Immortals themselves might descend into the garden, thus confirming not only the beauty of the Emperor's creation, but the virtue of the dynasty that could have produced it. Confidently Huizong named his great rockery Gen Yue, the 'Immovable Peak'.

The Influence of Geomancy

In fact the Gen Yue had been built not only as an exhibition of prosperity (and thus of virtue), but also as a practical means of achieving it. In China, at least since Tang times, it was widely believed that the fortunes of men were intimately affected by the forms of the landscape in which they lived. Good forces ran through the earth in streams or currents. Those who placed themselves well in relation to these currents naturally prospered, while those who allowed evil forces to concentrate around them fared badly. When Huizong came to the imperial throne he had no sons, and although he was only twenty-six years old this evidently caused him some distress. He ordered the *fengshui,* or 'wind and water' experts, to analyse the geomantic aspects of the imperial city. They diagnosed a case of too much flatness, maintaining that the Emperor could not have male heirs while the land to the north-east of his capital lacked height. Thus, building the mountain Gen Yue was a means of concentrating good forces in the landscape, and of blocking out those evil ones which prevented the conception of male children. By all accounts it was splendidly effective. However, although Huizong succeeded in fathering an heir, he lost his Empire in the process: building Gen Yue and stocking it with rarities from every corner of the imperial domain bankrupted the dynasty.

A notorious commissioner was in charge of collecting plants and stones: Zhu Mian from Suzhou, the garden city near Lake Tai where the finest water-modelled stones were found. At the time, it was said, no single item in the Empire was more expensive or difficult to come by than one of these stones. Zhu Mian was particularly good at locating them, often enough in the gardens of the Emperor's subjects, who were then encouraged to donate them to the Emperor as gifts. Huizong was certainly not the first Emperor to stock his garden in this way, but in this case a large number of rare items also found their way into 'The Green Water Garden', Zhu Mian's own private estate, which rivalled the Imperial park, greatly annoying the donors of the stones.

The expense of transporting enormous boulders was as heavy as the cost of buying them. The monk Zixiu, who wrote an account of the garden, believed that they were 'transported by some magic', but in fact barges carrying rocks blocked the nation's canals for days on end, disrupting the essential traffic of foodstuffs and raw materials. Huizong himself perhaps did not know of either the rapacious corruption of his commissioners or the appalling cost and disorder caused by building his paradise. He actually says quite blandly that 'the hard labour of transporting earth in baskets was not felt, and the sound of hammer and axe not heard'. Later, according to the official Song history, he did grow worried at the appalling expense, but the court eunuchs, avid for promotion and financial gain, were forever wanting to build.

The monk Zixiu witnessed the terrible consequences of Huizong's negligence and weakness:

Pergola on causeway, Southwestern shore, West Lake (Xi Hu), Hangzhou.

> In the second month of [1126] the capital was sacked. Its inhabitants all ran with me and took refuge on top of Gen Yue. Just then there was a heavy fall of snow. All the hills, valleys and trees looked as brilliant and beautiful as if they were part of a painting of all the most beautiful scenery that ever existed in the world. I toured round it for several days [*sic*], amazed at its unsurpassed beauty. Next year, in the spring, I returned again to visit [it], but it had been destroyed by the people. They hated the names, records, poems, written by the great ministers, and they broke down the tablets and threw them in the ditches. As to the beautiful trees and bamboos, the various buildings, people soon cut them down to use as fuel. The only thing that remained in the universe was the mountain itself, Gen Yue.[13]

'Alas,' he wrote, concluding his account, 'people in this world have long heard about Gen Yue, but who has seen it with his own eyes? And with what map or record may later generations make a study of it?' Imperial gardens prove just as transient as those of ordinary men; and of Huizong's great mountain there remains today no trace. Huizong himself ended his days as a prisoner of the Jurchen barbarians, and died in a tent among the Manchurian forests.

Hangzhou: the City on West Lake

The city of Hangzhou, which now became the capital for survivors of the catastrophe, has been almost universally acknowledged as one of the most seductive spots on earth. It was a city of canals, lying between a great river and a manmade lake, and surrounded to the west and south by wooded hills on which were situated some of the most famous monasteries in China. The exceptional beauty of the lake, with its willow causeways and islands and its rim of hills, its softly drifting mists and reflected pavilions, proved to be a far stronger influence on the new Emperor than the recent lesson of Huizong's downfall. Moreover, the southern states which now made up the Song Empire were exceptionally prosperous. It was a land of rice, salt, sugar and silk. Merchants and tradesmen grew immensely rich, and both the borders of West Lake and the main boulevard of the city were lined with great houses, villas and gardens. The impetus for such lavish display came inevitably from the court. With the re-establishment of the dynasty, many great artists of the old order began to find their way to the southern capital, and as soon as the court was settled, the Emperor began building palaces and pavilions all round the city and lake. Some fifty years later, when the dynasty had fallen, Marco Polo described this 'palace of the fugitive king' as 'the most beautiful and splendid . . . in the world'[14] – but of course he knew nothing of those earlier imperial seats which had already set the style.

His account sounds familiar enough, but the Hangzhou gardens were remarkable for their exceptional elegance and workmanship, and for their unparalleled setting. For one annual festival the entire park was hung with silk flags which shimmered in the breeze and were reflected in the water of the lake. In these delicious surroundings the Song Emperors, who in any case had little power over the great families of the south, dedicated themselves to pleasure. They frequently abdicated, spending the balance of their days surrounded by artists and poets in a round of banquets and boating and garden parties. Marco Polo drew the classic moral from the story of the last of these rulers: through 'his unmanliness and self-indulgence' he had forfeited 'all his state to his utter shame'.

Barbarian Splendour

Once again barbarians from the North took over; this time the Mongols, who now became the first barbarians ever to control the entire country, north and south. Less than a hundred years earlier they had been the most backward people in Asia, living lives of spectacular discomfort in the scorching heat and perishing cold of the great Mongolian wastes. They were herdsmen and nomads, with little use for agriculture, and even less concern for human life. When Genghis Khan first conquered Peking in 1215 the orgy of destruction lasted a month. Luckily it took over seventy-five years to conquer the rest of China, during which time the Mongols gradually became more civilized. When Genghis' grandson Kubilai finally took Hangzhou, he did not destroy the palaces of the southern Song, but merely allowed

them to crumble into ruins. Even more significantly he also began to feel the urge to build. He moved his capital from Karakorum, in Mongolia, to Peking inside the Great Wall, and there began to construct a great new city, imperial in its extravagance but with a certain barbaric splendour which sets it apart from the projects of other Emperors. In true imperial style he also began to build gardens and enclose great hunting parks. The previous rulers of Peking (the Jurchen Tartars who seized north China from the Song), had left a marshy lake in the city, fed by a stream from the Western Hills. This Kubilai enlarged and, with the excavated earth, added to a small island there until it was over a mile in circumference. Then he had it planted with evergreens and rare trees. Like Huizong's before him, Kubilai's hill was also embellished with rocks, but instead of the grotesque water-worn limestone of more recent Chinese taste, Kubilai's were all of lapis lazuli, like the rocky islands Emperor Ming Huang made for Yang Guifei. This, wrote Marco Polo, 'is intensely green, so that trees and rock alike are as green as green can be and there is no other colour to be seen'.[15] On top of it all Kubilai commanded a palace to be built which also was entirely painted and decorated in green. Later the garden would also contain two circular palaces rising out of the lakes, entirely covered in decorations of crystal so that to visitors they looked like palaces of ice. Walls within walls surrounded the new city and, between them, reminding the Mongols perhaps of their homelands, were great grassy parks planted with trees which had been

The Dagoba on Bai Ta Shan (White Tower Hill) in Beihai Park, Peking, just visible across roofs from within the Imperial City.

transported there on the backs of elephants.

Kubilai wintered in the city, but during the summer months he stayed in the north at his Mongolian retreat of Shangdu – the Xanadu of Coleridge. Like its Chinese equivalent, Shangdu contained a great palace made of marble on the outside and gilded within. To the north, enclosed by an extension of the city wall, lay a great hunting park accessible only from the palace. Here, in a dense grove of trees, a movable pleasure dome would be set up for the duration of Kubilai's stay, its sides and roof entirely made of split bamboos nailed together and supported by two hundred guy ropes of silk. Round about, with a hunting leopard on the crupper of his horse, the Mongol ruler hunted game to feed his famous mews of ten thousand goshawks, gerfalcons, peregrines and sakus.

The Last Great Imperial Gardens

In Peking there still exist today three great imperial lakes that date from Kubilai's time. These are in the Parks of the Sea Palaces, strung out along the western side of the Forbidden City in the very centre of the capital. Although the great Khan's lapis lazuli rocks, green pavilion, and baths and fountains no longer exist, the Qionghua Island in today's Beihai Park is the very same hill of excavated earth that Kubilai knew, for although the first Emperor of the Ming destroyed all Kubilai's works, when the Yongle Emperor later decided to rebuild Peking as his own capital, he made use not only of Kubilai's street and palace foundations, but also of his lakes and islands.

Compared to the rulers of the foreign dynasties which both preceded and followed them, the Ming Emperors did not much care for extensive travelling, and preferred their retreats as close as possible to the capital. Thus Kubilai's lake shores were built up as summer residences for Yongle and his successors. The lakes themselves were also further excavated, giving the three Seas the shape they have today, and they were embellished with rich plantings of trees and a water line of irregular rocks. After all, in the Chinese scheme of things, these lakes with their pavilions half hidden in trees were a 'natural' and essential balance to the rectangular palaces beside them, even if they were in fact entirely artificial.

Later the great garden builders of the Qing dynasty also repaired and added to these gardens, so easily accessible from their winter quarters. By the end of the dynasty in 1911 they were a mass of palaces and pavilions, bowered in ancient trees and crowned by a great white tower, the Dagoba, shaped rather like a hand bell or a fat-bellied bottle.

In the 1930s these 'Parks of the Sea Palaces' were visited and later evocatively described by two foreigners who particularly appreciated Chinese gardens. Both George Kates and Osvald Sirén spent many years in Peking and innumerable hours in these parks, and both agreed that they were the loveliest and best preserved of all the traditional pleasure grounds in China. At that time, although virtually no buildings remained from the Ming period (1368–1644), the atmosphere of unbroken peace was so strong that even ill-proportioned twentieth-century additions did not detract from the total effect. The parks were so large, their buildings so complex and so many, and their planting so rich, that it was possible to explore for months and still not be familiar with all their features. And although the parks were open to the public (bearing a similar relation to Peking as Hyde Park and the Serpentine do to London), there were still secret corners that remained deserted all day long. Such a place was the Ying Tai or Ocean Terrace, an Island of the Immortals some four hundred and fifty yards in circumference, rising in rocky caverns from the waters of the Nan Hai, or Southern Sea. Sirén's photographs of this island invest it with a strange melancholy. Here, in 1898, the old Empress Dowager had imprisoned her son, the legitimate Emperor Guangxu, who had encouraged a *coup d'état* against her rule. For two years he remained a captive on the island, mocked in his helplessness by the flowery names of the rooms that were part of his prison. Perhaps he paced round the clustered halls and terraces, plotting once more to bring down the malignant power of the Empress; or perhaps, as Sirén suggests, he gave up the struggle and gradually faded away, 'like the grey daylight over the still lake'. What is certain is that just beyond the high walls of his prison, life in his capital went on totally unaffected by his fate. There he remained, alone at the centre of a vast and teeming Empire, seeing only lake and sky and the sun rising and setting, as if he and his tiny entourage were the only people on earth.

Gardens as 'Refreshment for the Heart'

This of course, was exactly the effect the great imperial garden makers had aimed to create. The encircling walls of a garden were designed to cut out the world of men and transport imperial thoughts on to a plane quite distinct from the political. When the Qianlong Emperor felt an urge to build gardens, he argued that a retreat was essential for his moral health:

> Every Emperor and ruler, when he has retired from audience, and has finished his public duties, must have a garden in which he may stroll, look around and relax his heart. If he has a suitable place for this it will refresh his mind and regulate his emotions, but if he has not, he will become engrossed in sensual pleasures and lose his will power.[16]

Thus the paradox: for wise and conscientious rulers, a garden is an essential place of spiritual rejuvenation; yet, as we have already seen, it can also be a temptation to sensual excess. For the Qianlong Emperor it was both – not that he was a weak monarch, or unconcerned with his duties, but because he was swept away with a passion for the beauties of nature and with the desire to create beautiful places. At the same time he was acutely conscious of the frugal example left him by his grandfather, the illustrious Kangxi Emperor, in whose lovely but quite modest garden he had lived as a child.

When he first came to the throne, the Qianlong Emperor was torn between the longing to landscape and the very serious desire not to 'bring shame upon myself and my ancestors' by any form of excess. With what was clearly a Herculean effort of will, he held back for three years during the period of mourning for his father's death, continuing to live modestly in his predecessor's apartments and manfully rejecting all the court eunuchs' suggestions for building. All this is known, however, from an apology the Emperor wrote to justify his first building programme. For even Mencius would have found it hard to restrain the Qianlong Emperor for long.

Eventually, on the grounds to the north of his father's old garden, the Chang Chun Yuan or Garden of Expansive Spring, a thousand workmen could be seen forming hills and valleys and lakes and winding canals, with all the pavilions, hills, bridges and rockeries appropriate for an imperial retreat equal to the most lavish in history. The style of landscaping could hardly be described as very innovative, yet as the last great flowering of the imperial tradition it was perhaps fitting that it should have been this garden – the Yuan Ming Yuan – that made such an impact on Europe.

It was a garden that included all the themes we have seen develop: a landscape of 'nature improved', lavish pavilions, great halls for imperial entertainments and audiences, and tiny corners for study. In addition the Emperor built in it a vast library to house one complete set

RIGHT *The Da Shui Fa, or Great Fountains, Yuan Ming Yuan, near Peking, engraving of 1786. These fountains were part of the European follies which the Qianlong Emperor commissioned from the Jesuit Father Benoit in the Northwestern corner of the Yuan Ming Yuan.*

BELOW *Remains of the south façade of the Da Shui Fa as they are today.*

PAGES 74–5 *Brilliant colours in a newly restored pavilion, Peking. During the long northern winters such pavilions decorate the gardens of the capital like flowers. In summer they are half hidden by foliage.*

of the great Qing collection of classical works. There were also temples, the complete reproduction of a southern city street and of rustic farms, drill grounds for training troops and, as Father Attiret noted, 'little parks for the chase' – although, judging by the plans of the garden, any hunts staged there must have been on a very small scale. Inevitably the garden also contained an immense collection of different plants and many small menageries in which interesting or unusual beasts were kept. Most extraordinary of all – and this was certainly an innovation – were a series of buildings in carved stonework occupying the north-east boundary of the garden. Here the Qianlong Emperor commissioned the Jesuits to build him a collection of garden halls with fountains and a maze in imitation of European baroque – a kind of reverse chinoiserie. Of all the treasures of the Yuan Ming Yuan, the ruins of these are all that remain today.

All this the Qianlong Emperor built while continuing to affirm his frugal principles. One justification he suggested was that with all these lovely gardens at their disposal, his descendants would be 'spared' [sic] any further expense of building. Nevertheless, when he finally completed the Yuan Ming Yuan, he wrote a memorial promising that this was the end, and no doubt he genuinely convinced himself that it was. But how could it have been? The Emperor had only to see a charming hill or stretch of water to be seized with the desire to improve it. Beyond the Great Wall at Jehol, where the Manchu rulers kept great hunting preserves, he added lakes and temples to make another spectacular garden, while in the foothills of the Western Hills near Peking, he embellished a series of wilder gardens with pagodas and pavilions.

One site here might have tempted even someone without the Emperor's obsession. It was a relatively large, naturally formed hill which stood alone to the west of the Yuan Ming Yuan with the shifting colours of the Western Hills lying behind it like a screen. Here the waters of the Jade Fountain, which flow eventually into the Sea Palace parks in Peking, had been dammed to form a lake almost four miles in circumference.

At first the Emperor used the lake as a practice ground for sea battles, but soon he realized its potential as a garden: 'When I had this beautiful hill and this beautiful lake,' he said, protesting as usual a little too much, 'how could I refrain from building the terrace and buildings suitable to it?'[17] The year 1750 gave him a perfect excuse: for his mother's sixtieth birthday he transformed the place into a garden with bridged and willowed causeways in imitation of Hangzhou, a city which she specially loved. Thus 'filial piety', not 'extravagance', added another huge enclosure to the Emperor's estates.

Just over a hundred years later, after the destruction of the gardens in 1860, exactly the same reason was used as an excuse to repair it. This time the occasion was the sixtieth birthday of the Dowager Empress Cixi. The Empire by then was tottering and the imperial coffers relatively empty, but gifts of cash came pouring in from rich merchants seeking honours. In addition a substantial loan was raised, partly abroad, to equip and modernize a Chinese navy. To appropriate this sum, the old Dowager declared she would set up a naval academy on the lake, so reviving the Qianlong Emperor's original justification for creating it. However, what she actually built with the money were the courtyards and halls of the garden which can still be seen there today. The only ship the Navy got, goes an old joke, was the Empress's marble Tea House which, shaped like a Mississippi paddle steamer with an Italian loggia perched on top, marks the end of the garden's lakeside galleries.

As George Kates notes, the Dowager Empress's affection for the place has left on it 'the dubious stamp of her time'.[18] The halls and courtyards on the eastern shore are, on the whole, crowded

RIGHT *Looking south-east towards Peking from the top of Wan Shou Shan (Longevity Hill), Yi He Huan (Summer Palace). In the distance the wide expanse of Kunming lake, edged with willow causeways, surrounds. the Island of the Dragon King reached by the Seventeen Arch bridge.*

RIGHT, BELOW *A water-modelled rock, set up like sculpture on a marble stand, marks the entrance to the Yu Hua Yuan, the back garden of the Imperial City, Peking. See page 137 for a plan of this strange garden.*

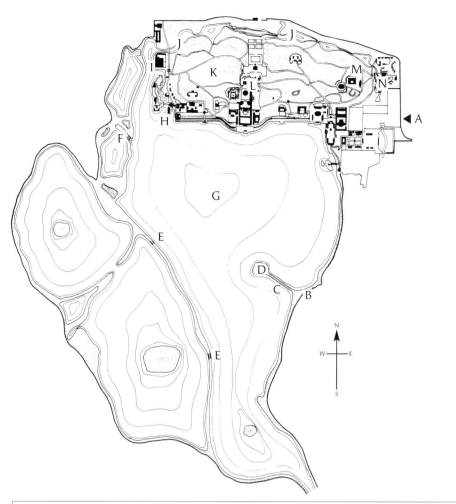

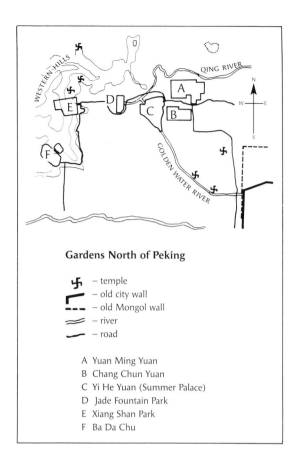

Gardens North of Peking

卐 – temple
⌐ – old city wall
--- – old Mongol wall
≈ – river
— – road

A Yuan Ming Yuan
B Chang Chun Yuan
C Yi He Yuan (Summer Palace)
D Jade Fountain Park
E Xiang Shan Park
F Ba Da Chu

Yi He Yuan

A Entrance	F Jade Girdle bridge	K Longevity Hill (Wan Shou Shan)
B Bronze ox	G Kunming lake	L Hall of incense to the Buddha
C Seventeen arch bridge	H Marble boat	M Pavilion of Joy in Agriculture
D Island of the Dragon King	I Imperial boat houses	N Xie Qu Yuan
E Bridges on causeway	J Back lakes	

ABOVE, LEFT   *Plan of Yi He Huan (Summer Palace). (See colour, page 77.)*

ABOVE, RIGHT   *Plan of gardens north of Peking.*

OPPOSITE:
ABOVE   *Marble Boat, Yi He Yuan, (Summer Palace), near Peking. The Dowager Empress Cixi added this tea house on to a jetty built originally for the Qianlong Emperor in the eighteenth century. It stands at the end of the mile-long gallery that runs along the northern shore of the Kunming lake.*

BELOW   *For actual boating expeditions on the lake the Dowager Empress also had a real, steam-powered paddle boat, which was rescued from the bottom of the lake during dredging some years back. It is now on view just behind the marble boat.*

and heavy, the rock-work seldom inspired, and high on its massive terrace, dominating the whole mountain and the lake and gardens below, the great octagonal tower (or 'Hall of Incense to the Buddha'), has always seemed to me an oddly oppressive and lumpish piece of work.

Nevertheless, by some extraordinary magic, the Yi He Yuan remains an irresistible place and one full of surprises. North-east of the entrance there still exists a secret pool encircled by pavilions and enclosed within its own walls, called the 'Garden of Harmonious Interest', which the Qianlong Emperor built in imitation of a famous old garden in Wuxi. It is surrounded by a low artificial mound which seals it off from the world outside: a garden within the garden.

Another source of delight is the play of contrasts, which takes place on a massive as well as an intimate scale. For example, the great expanse of the Kunming lake, spreading out south below Longevity Hill, is balanced to the north by the steep shoreline of the Back Lakes which zig-zag through a deep and shady gorge crossed by arched stone bridges. The eastern maze of courtyards is counter-balanced by country walks winding up the hills beyond them, through groves of lilac trees which turn the whole place fragrant in Spring. And the water itself is

balanced not only by the rocky island of the Dragon King and by Longevity Hill, but also by all the range of the Western Hills beyond.

It is these that really make the garden – nature 'borrowed' from without, as the Chinese say. Through the days their reflected colours change from misty greys to sharply outlined flanks of amethyst and blue. Sometimes they seem close enough to touch, at others as insubstantial as a veil. As usual in Chinese gardens it is the effects of light and seasonal change that here have been celebrated with such grace and grandeur, so that from time to time even foreign barbarians have felt themselves to be among the dwellings of the Immortals.

Some years after the burning of the Yuan Ming Yuan a Chinese minister to London reassessed the loss, and added to the traditional moral a new awareness of China's place in the world. For him the burning signified his country's awakening:

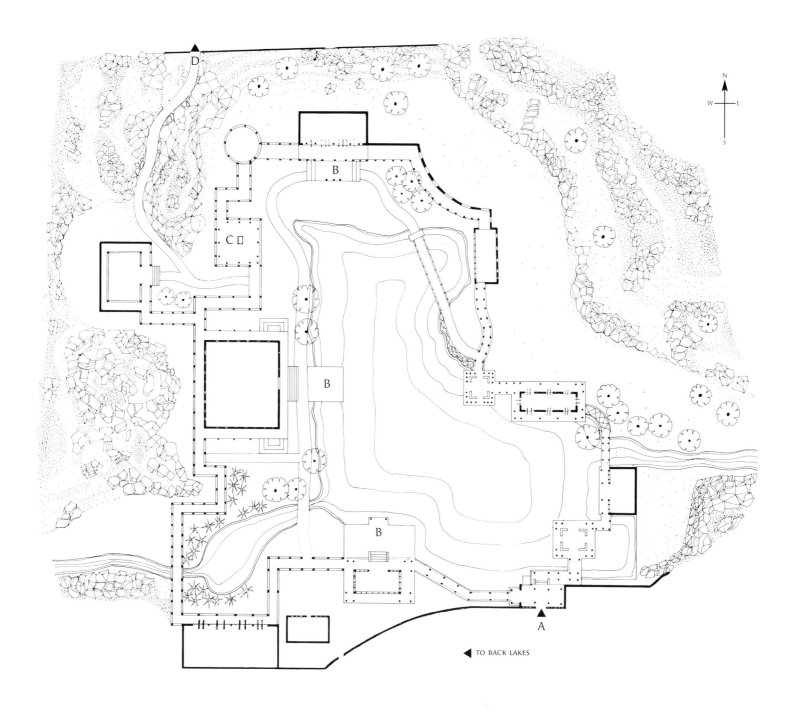

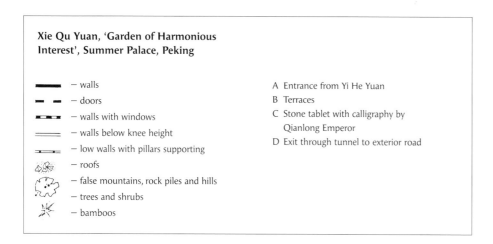

Xie Qu Yuan, 'Garden of Harmonious Interest', Summer Palace, Peking

▬▬▬ – walls

▬ ▬ ▬ – doors

▬◘▬◘▬ – walls with windows

═══ – walls below knee height

╤╤╤ – low walls with pillars supporting

✿ – roofs

❀ – false mountains, rock piles and hills

– trees and shrubs

✳ – bamboos

A Entrance from Yi He Yuan
B Terraces
C Stone tablet with calligraphy by
   Qianlong Emperor
D Exit through tunnel to exterior road

Plan of the Xie Qu Yuan (Garden of Harmonious Interest), a 'garden within a garden' in the Summer Palace, Peking, here compared to plan of the Ji Chang Yuan, Wuxi. The Qianlong Emperor commissioned the Garden of Harmonious Interest after he had visited this famous old garden in the south. It is not meant to be an exact copy, but to capture the spirit of the original. The northern garden is much stiffer and more formal than the other, although in its own way equally lovely. (See colour, page 36.)

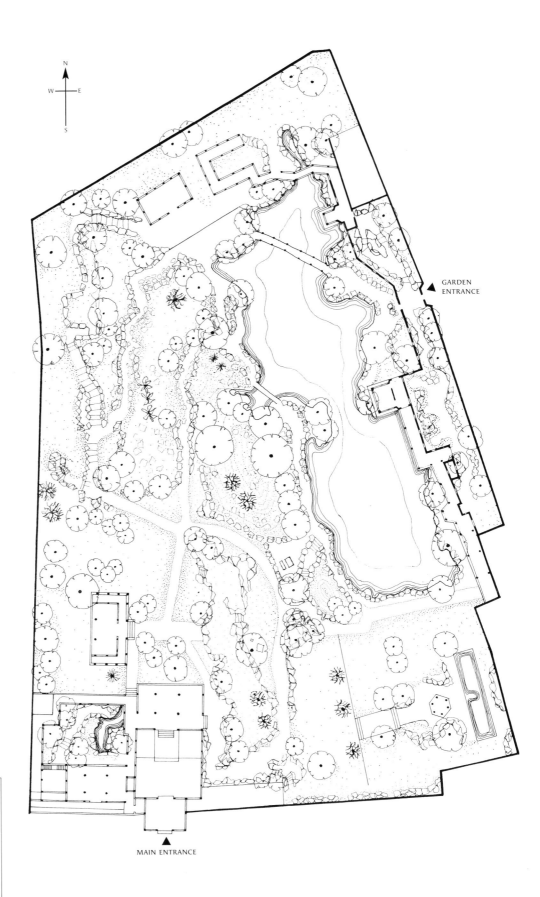

GARDEN
ENTRANCE

Ji Chang Yuan, Wuxi

— walls

— walls with windows

— walls below knee height

— walls with pillars supporting roof

• • — pillars supporting roof

— rock piles and hills

— bamboos

— trees and shrubs

MAIN ENTRANCE

By the light of the burning palaces which had been the delight of her Emperors, [China] commenced to see that she had been asleep while all the world was up and doing. . . . The Summer Palace with all its wealth of art was a high price to pay for the lesson we there received, but not too high, if it has taught us to repair and triply fortify our battered armour: – and it has done so.[19]

Perhaps he was right in the long run, yet, as we have seen, the Empress Dowager's reaction, typically enough, was to rebuild the gardens with money raised for her Empire's navy: when China needed a navy to fight against Japan, she thus had none available. As Mencius pointed out succinctly long ago, the extravagances of rulers fall heavily on the people, and yet the urge to build beautiful places is not only a selfish one. Despite all the examples of indulgence down through history the Chinese always seem to have kept an alternative set of values with which to judge their Emperors' works. It is of course a Taoist alternative, here expressed by an old gardener as he looked across at the tiled roofs of the Summer Palace glimmering in the sun:

Many people nowadays blame the Empress for using the money to build a garden instead of buying warships to fight with foreign countries. But I think she did a wise thing. War is a beastly idea after all. Look! . . . Who would ever have thought that such beautiful things could be made by the hands of men, had she not spent the money on it?[20]

ABOVE *Back Lakes, Yi He Yuan (Summer Palace), Peking. In contrast to the wide expanse of Kunming lake to the south, the northern slopes of the hill fall steeply down to a narrow canal, sharply edged in cut stone. Here the Qianlong Emperor staged naval battles, and built the replica of a Suzhou street, so his ladies could enjoy buying things 'like ordinary people'. These Back Lakes lead round eventually to the Xie Qu Yuan (Garden of Harmonious Interest).*

RIGHT, ABOVE *Xie Qu Yuan, Yi He Huan (Summer Palace), near Peking. Hidden behind walls, pavilions linked by open galleries surround the lake in this 'secret' garden within the garden.*

RIGHT, BELOW *Yi He Yuan (Summer Palace), near Peking. Looking from Longevity Hill to the pagoda on Jade Spring Hill and the Western Hills beyond.*

THE GARDENS OF THE LITERATI

To the art of making gardens the Emperors of China contributed two major themes: the vision of the magical dwellings of the Immortals, and the idea of the park as a microcosm in which might be found all the riches of the world. Both these ideas had also found expression in private gardens during the Han dynasty – such as in that famous park with its artificial mountain that cost Yuan Guanghan his head. With the end of this dynasty, however, in the third century AD, private gardens began to show the influence of changing values, and to develop two quite different themes that were also to become essential aspects of the Chinese garden. One concerns the expression of Taoist nature philosophy in gardens; the other the growth of a rich, unique and fascinating connection between the arts of gardening, painting, poetry and calligraphy.

The Effect of Taoism

Taoism was hardly a new philosophy at the end of the Han dynasty: its 'great books', the *Dao De Jing* and the *Zhuangzi* date back to the third and fourth centuries BC, while the philosopher Lao Zi himself (the supposed author of the *Dao De Jing*) is thought of as a contemporary of Confucius (551–471 BC). Moreover, the various Chinese views of nature – as an awe-inspiring wilderness, a place of magic, a provider of sustenance – had also long been accepted, while, as we have seen, the search for an elixir of immortality had found many followers at the imperial courts. With the collapse of the Han, however, and the corresponding loss of faith in the old order, all these disparate elements began to synthesize in a new way. The terror of new northern invaders, political confusion, hypocrisy and weakness at court: all these led good men to a loss of faith in Confucian values. They could no longer believe in an ethical system that had led China so far astray, and they began to look at alternative philosophies. Mentally they escaped from the chaos into a new interest in Taoist views of nature, and later also into Buddhism. Physically many of them fled south to lands which were still, to them, scarcely civilized.

The landscape in which they found themselves was sparsely populated, wild and moist. Its valleys, which fell away to a tangled skein of lakes and marshes, were gentler than those of the north, and its deeply folded mountains were softer and more lush. Here in the mid-third century Seven Sages of the Bamboo Grove sat discussing the inner points of esoteric Taoism. When approached by vulgar men who talked of worldly profits, they rolled their eyes up into their heads and waited until the vulgarians departed. One of these sages, Liu Ling (*c.* AD 221–300), upset people who came to see him by sitting naked in his room. When they complained he answered: 'I take the whole universe as my house and my own room as my clothing. Why then do you enter here into my trousers?'[1] Although he no doubt enjoyed such opportunities for wit, Liu Ling was also expressing a fundamental Taoist view. He had achieved unity with all existing things, however great or small; his trousers, his house and the universe were totally interchangeable.

It is perhaps because of this sense of unity with nature that the appreciation of wild landscape, and the urge to paint or re-create it, occurs in China some eight centuries before its emergence in the West. Philosophical Taoism in particular stressed the oneness and continuity of the sensual, material and spiritual worlds, but all the Chinese schools also held that everything in the

Universe was made up of the same basic material. This they called the *qi*, 'breath' or 'ether', a notion that is fundamental to the Chinese view of life, and particularly to fields of activity such as landscape painting.

For the Chinese 'The pure and light *qi* rose to become heaven, the muddy and heavy fell to become earth; the breath which harmoniously blended both became man'.[2] Thus, man stood midway between heaven and earth, united by this essence with everything else in existence. One point of a garden, evident in so many paintings, is to trigger this latent sense of unity with the universe. When a man stands in the spring snow, oblivious of the cold, and gazes into a cloud of plum flowers, he feels as if his heart is suddenly flowering too in sympathy with the tree – as vigorously and spontaneously as the blossoms themselves. In this way he experiences, through his garden, the mysterious workings of the Dao ('Tao' in an earlier system of romanization, hence 'Taoist').

The Dao of the Taoists

In China all the philosophical schools used the word *dao* (meaning literally 'way'), by which they referred to the method of proper conduct each considered the right one if the world was to run harmoniously. For the thinkers known as Taoists, however, this word came to symbolize 'the totality of all things whatsoever':[3] it was more than merely a 'way' to follow (however successful), it actually *was* the whole process. The Dao was the sum of everything, past, present and future, in its state of eternal transformation. Like a great seamless web of all creation, it formed and dissolved quickly, as in the patterns of clouds, or over aeons of time, like the earth itself. Impersonal, the Dao never ceased to change, yet never repeated itself. And in a larger sense it remained unchanging, since the sum of its forces was always the same.

Exquisite Jade Rock, Yu Yuan, Shanghai. One of the finest Taihu rocks still to be seen in Chinese gardens, this monolith was reportedly commandeered for the Emperor Huizong's great rock garden in the eleventh century. However, it was so difficult to transport that it was put on one side and forgotten.

Woodcut from the Hong Xue Yin Yuan Tu Ji *by Lin Qing. The inspector-general sits in his study beside a magnificent Taihu rock in his courtyard.*

In gardens the concept of the Dao – always purposely vague and ill-defined (since any definition would impose limits on it), was symbolized by huge, strange, standing stones which the Chinese place (alone or sometimes in groups), somewhat as we do sculpture. The stones, billowing out from narrow bases, hollowed by weather and time, pitted with holes and seemingly frozen in perpetual motion, are exceptionally powerful images. Moreover, they are not just symbols of the Dao but, since they too are part of the web of existence and subject to the inevitable processes of time and decay, actually part of the Dao they represent.

Taoism in Gardens

Apart from such specific images, Taoism is also of great general importance for Chinese gardens. Obviously the cyclical idea of life in death, of autumn implying spring, is suggested by the annual blooming and death of flowers in any garden, but in China Taoism gave it added force and meaning:

Since death and life . . . attend upon one another (says the *Zhuangzi*), why should I account either an evil? Life is accounted beautiful because it is spirit-like and wonderful. Death is accounted hateful because it is foetid and putrid. But the foetid and putrid, returning, is transformed again into the spirit-like and wonderful; and then the reverse occurs again once more. Therefore it is said that all through the universe there is one *qi*, and therefore the sages prized that unity.[14]

The Taoist aim was to become one with these eternally shifting patterns, thus overcoming fear and all uncertainty. One story explains the idea: Confucius was taking a walk with his disciples along the bank below the Lüliang waterfall. In the raging torrent, where even a turtle would not have survived, they saw a man's head bobbing in the waves. The disciples dashed forward to help, but the man climbed out quite unconcerned. 'Have you a *dao*,' Confucius asked, 'to tread in the water like this?' 'No,' the man replied (nevertheless giving a grandiloquent explanation that reveals he is a Taoist): 'I began in what is native to me, grew up in what is natural to me, matured by trusting destiny. I enter the vortex with the inflow and leave with the outflow, follow the Way of the water instead of imposing a course of my own; this is how I tread it.'[15]

Designers of gardens might boast of similar intuitive powers, by which they follow nature's inherent lines to make the most of an existing site. The art of gardening, however, cannot be specified and prescribed any more exactly than the true Way. The Dao is suggested through aphorism and contradiction: to gain, you must yield; to grasp, let go; to win, lose. And as we will see, Chinese gardens are designed with the aid of a similar set of contrasts: the method for teaching garden building equally involves evocation and suggestion rather than a precise formula.

A further main idea which Taoism contributed to garden art grows out of this intuitive, contradictory approach. This is the idea of *wu wei,* which means non-action, or at least no action contrary to nature.[6] A garden reflects the principle of *wu wei* because all its elements are designed to flow with the soft curves of nature; and also because in creating his garden a man is supposedly free to express his own unfettered imagination. Indeed, gardens are 'havens of inner strength',[7] where a man may harmonize with the inevitable passage of seasons, while the beauty of the physical world makes the struggles of everyday life seem less important. Thus, *wu wei* is also an ultimate justification for retiring or retreating from society into a passive state – like an enlightened sage, 'ignorant', pure, and blank as an 'uncarved block'.[8]

Extravagance in the Taoist Mould

Despite the Taoists' insistence on a simple and total unity with the Dao, the gradual, general acceptance of these ideas did not necessarily prevent the creation of elaborate gardens. Although seriously involved Taoists no doubt lived simply in natural surroundings, like those of Liu Ling and the Bamboo Grove's Seven Sages, other men who professed ideas in sympathy with Taoist thought still managed to make themselves extremely comfortable at great expense. Such a man was the third century Duke Shi Chong, who made a garden called the 'Villa of the Golden Valley', that became a byword for all that was most lavish. For a party Shi Chong once constructed an avenue of brocaded screens within the garden that was fifty miles long – and that only to trump a rival whose similar avenue stretched merely forty miles. Another time he made his concubines walk along a path dusted with aloes powder (exceptionally rare and costly at the time), just for the pleasure of ordering those whose feet left more than the whisper of an imprint to starve until they were as light as thistledown. He does not sound like a man sympathetic to Taoist ideas, yet from his own description of the estate he might seem like the simplest and least ostentatious of men. 'When I was fifty,' he wrote,

> I had to give up my post (at court) because of an unfortunate occurrence. . . . The older I became
> the more I appreciated the freedom I acquired; and as I loved forest and plain, I returned to my villa.
> . . . In front the prospect extended over a clear canal. All around grew countless cypresses, and
> flowing water meandered around the house. There were pools there, and an outlook tower. I bred
> birds and fishes. When I came home I enjoyed playing the lute and reading.[9]

In other words this garden, famous down the ages as one of the most lavish in China, was to its owner a simple pastoral retreat. Perhaps it was; or perhaps Shi Chong was playing safe, minimizing his riches to avoid the envy of others. Or perhaps, and this seems reasonable, he was simply expressing the increasingly accepted clichés of gentlemanly taste.

Landscape as a Spiritual Force

By the third century AD a man's character might be judged by the quality of his response to wild nature. For instance, an official at the time compared himself favourably to the Prime Minister by suggesting that although 'in official matters I am no better, in appreciation of mountains

'The Jin Gu (Golden Valley) garden of Shi Chong', by Qiu Ying, sixteenth century. Despite Shi Chong's protestations of modesty his third-century garden has gone down in history as one of the most opulent estates of all time. Here the artist has portrayed it in the style of an elaborate garden of the sixteenth century, with raised marble peony beds very like those that can still be seen in Peking's imperial gardens today.

I think I excel him'.[10] He said this to the Emperor, and had no need to say more, for a man who truly loved mountains was by then accepted as a man of deep spiritual power.

Xie Kun (AD 280–322), who thus described himself, belonged to a family with huge estates in Kuiji, one of the most dramatically beautiful areas of south-east China. Here, in the mid-fourth century, members of the Xie clan wandered with their friends across the hills and 'lived at their ease. They went out to fish and shoot among the mountains and rivers, and came home reciting verse. They had no worldly thoughts'.[11] Among their other-worldly thoughts, however, was the desire for longevity, and they searched the mountains for the herbs of immortality. Through these activities they began to look on the whole of nature as an *objet trouvé* and to build little pavilions from which to regard it. From this was born the notion of the 'borrowed landscape', in which the act of choosing a site, and building on it a little viewing place, turned the whole landscape into a 'garden'.

The first pavilion of this kind to become famous was the Orchid Pavilion, where one of the finest scrolls of calligraphy was composed in AD 353. In this scroll Wang Xizhi (321–379) records the spring festival of that year, when a literary group met in the pavilion for a traditional poetry contest, floating cups of wine down the nearby stream while they wrote elegant *vers d'occasion* under the trees. Those who could not finish a poem before their cup floated past them had to drink its contents as a forfeit, an outdoor literary game that became a favourite in gardens. It was even incorporated into the design of certain pavilions, which had special 'wine-cup streams' let into their floors.

The Literary Tradition

By the fifth century two great Chinese poets, Xie Lingyun and Tao Qian (also called Tao Yuanming), had already formulated Taoist feelings for nature in profoundly expressive verse.

'Gathering at the Orchid Pavilion' (Lan Ting), by Qian Gu, 1560 (detail). The celebrated calligraphic record of this occasion (which took place in AD 353) describes it as 'a clear day, when the sky was clear and the weather good. A gentle breeze blew softly. When we looked up we beheld the vast universe. When we looked around we could perceive the fullness of things.' The natural landscape, 'borrowed' by the pavilion, becomes a jardin trouvé, which in its turn is seen as a microcosm of the universe. (Translation by Roderick Whitfield.)

Later, their responses would become standard attitudes for the gentry, finding expression in both landscape painting and in gardens. But, just as importantly, both men also made gardens themselves, and wrote about them in their work, so setting in motion a long literary tradition that is an essential part of the Chinese garden. Xie Lingyun (AD 385–433), managed to combine in his life most of the contradictory aspects of his tumultuous age. He was a Duke as well as a poet, an intriguer at court and a Taoist herbalist, the owner of vast estates and a lay monk who was to become one of the foremost Buddhist thinkers of his day. He first went to a monastery perhaps for physical safety, following a debacle at court, but there discovered solace and stimulation in Buddhist doctrine that was to hold him all his life.

Buddhism in Gardens

Buddhism, which began to filter into China as early as the first century BC, did not require that one should give up all previous beliefs. In fact it took over many Taoist expressions, including the term Dao itself, partly in order to be more comprehensible to the Chinese. Also, in China it took over and adapted the Taoist attitude to nature, transforming the vast misty landscape into a symbol of the void. The Buddhist notion of the universe centring around a stupendous central peak (Mount Meru or Sumeru) fitted in well with existing mountain worship; while Amitabha's paradise garden, with its emphasis on mountain ranges separated by oceans, mirrored the magic dwellings of the Immortals. In fact the effect of Buddhist imagery on Chinese gardens was to intensify and add new meanings to ideas and motifs which had already been absorbed. In one area, however, it did add an important new theme: instead of the Taoist adept wandering alone in the hills, the Buddhists often joined together to enjoy the fruits of solitude. To the single hermit, then, was added the idea of the mountain monastery.

It was to such a place that Xie Lingyun first retired; a great monastery founded by his contemporary, the monk Huiyuan, on the holy mountain of Lu Shan. The place had been chosen as a Buddhist retreat because of its natural beauty and an atmosphere so deeply peaceful that the monks felt it would help in the attainment of Nirvana. Many laymen and monks gathered there in the late fourth century, and when they returned home after their stay they took with them the memory of its magnificent landscape which they often tried to re-create in the form of 'Lu Shan Parks'.

When Xie Lingyun next had to leave court he went to his own southern estate at Kuiji, and tried to continue there this same connection with nature and Buddhism. At the same time he busied himself with many great improvement schemes, building a road through the forest for a hundred miles, draining lakes, planting trees and making gardens. For walking over his estate he devised shoes with hooks in the toes for going uphill which could be moved to the heels for coming down. His verses established many of the themes – such as ageing, immortality, and the purity of mountain solitude – which were to affect later garden designers:

> By tall trees a century old,
> I breathe the fragrance of a myriad ages,
> I drink the springs of antiquity . . .
> Hemmed in by mountains, there seems no way out.
> The track gets lost among the thick bamboos.
> My visitors can never find the way,
> And when they leave forget the path they took.[12]

Such Taoist seclusion did not satisfy him all the time, however. To the Taoist vision of the hermit among the crags, Buddhism had now added the ideal of monastic life, in which a community of men sought spiritual enlightenment together. So, on Mount Stone-cliff, Xie Lingyun built a retreat for himself and two monks. Beside its courtyard a waterfall 'flew past the windows', and below lay acres of forest rising from misty valleys.

The Native Pool

Many men in office wrote longingly of getting back to the mountains, but hung on to their positions, putting aside their cash for the days when they would retire into nature. 'Hills and waters' nostalgia is a constant theme in Chinese poetry, and in the end takes on all the inconsequentiality of a conventional pose. In the hands of China's great poets, however, it still has a bite and freshness that makes the old conflict between social obligation and self-cultivation genuinely moving.

One of the first people to crystallize this emotion and make it accessible to quite ordinary people was the poet Tao Qian, who lived from AD 378 to 427. Unlike his contemporary Xie Lingyun, he was not the scion of a famous family and he had no private income. What he did have was a genuinely free spirit which caused him a lot of agony:

> When I was young I was out of tune with the herd;
> My only love was for the hills and mountains.
> Unwittingly I fell into the web of the world's dust
> And was not free until my thirtieth year.
> The migrant bird longs for the old wood;

'Cultivating chrysanthemums by a thatched lodge', by Gao Fenghan, 1727 (detail). The scholar working happily in his garden is directly reminiscent of Tao Qian, although the poet had died exactly 1,300 years earlier.

The fish in the tank thinks of its native pool,
I had rescued from wilderness a patch of the Southern Moor
And, still rustic, I returned to field and garden . . .
Long I lived checked by the bars of a cage;
Now I have turned again to Nature and Freedom.[13]

Tao Qian's problem, he later claimed in self-defence, was that his house was full of children, the rice jar empty, and 'I could not see any way to supply the necessities of life'. Friends and family nagged him to get a job, and at last he gave in and took a minor official post in a small town. Before a week was out he knew it was a disastrous mistake:

> Hunger and cold may be sharp but this going against myself really sickens me. Whenever I have been involved in official life I was mortgaging myself to my mouth and belly, and the realization of this greatly upset me. I was deeply ashamed that I had so compromised my principles.[14]

Finally, using the death of his sister as an excuse, he packed up and went back to his house. He found its fields and garden all overgrown with weeds and its three paths almost obliterated, and could hardly bear the joy of seeing the trees in his courtyard again:

> I lean on the south window and let my pride expand.
> I consider how easy it is to be content with a little space.
> Every day I stroll in the garden for pleasure.
> There is a gate there, but it is always shut.
> Cane in hand I walk and rest,
> Occasionally raising my head to gaze into the distance.
> The clouds aimlessly rise from the peaks.
> The birds weary of flying know it's time to come home;
> As the sun's rays grow dim and disappear from view,
> I walk around my lonely pine tree, stroking it.[15]

Unlike Xie Lingyun with his great estate, Tao Qian was forced to take up farming for a living. The life was tough. Once he lost everything in a fire, and at the best of times the family fare was simple and the rooms bare. Sometimes he wrote in despair – and yet at bottom he always knew that what he was doing was right for him. In the end his fears of a wasted life proved unnecessary, for his poems became some of the best known in all Chinese literature, and his simple cottage garden one of the models for every future recluse.

Tao Qian's name is most often associated with chrysanthemums, on account of an image in one of his best loved poems. This is a work which gives almost perfect expression to the Taoist-inspired sense of unity with nature. In it the poet captures that long moment at the end of the day when the light is low and clear, but dusk not far away. He is down at the

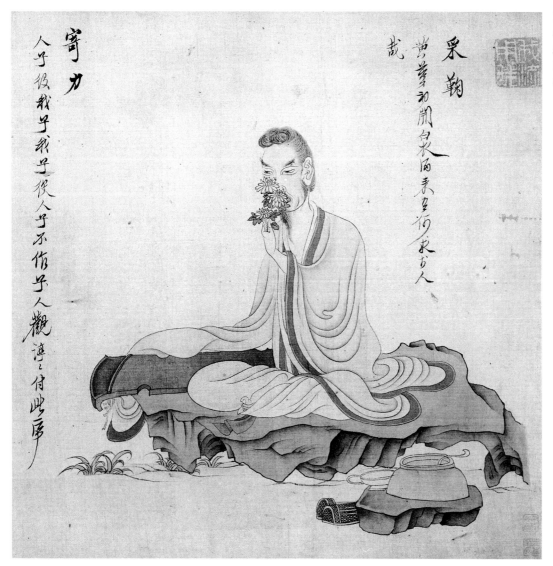

The poet Tao Qian with a chrysanthemum, detail from 'The Homecoming', a seventeenth-century scroll by Chen Hongshou. Note the flat-topped rocks used as a seat and table in the garden.

bottom of his garden by the eastern hedge, plucking chrysanthemum petals with which to make a life-prolonging infusion in wine. Elsewhere he has said how foolish such hopeful concoctions are, and as he straightens now, he suddenly catches sight of the light on the distant southern hills – the place where he will one day rest in his tomb. In the fresh air of evening the hills stand out clearly, as if they were just over the hedge. Above his head, birds in twos fly home from them to roost. At that moment Tao Qian seems as much part of the landscape as the hills in the distance, the birds, or the chrysanthemums he holds in his hands. The image of him standing there catches an eternal moment of suspended time when man and nature seem to be in perfect accord. 'In these things,' he says, 'there lies a deep meaning. Yet when we would express it, words suddenly fail us.'[16] It is this inexpressible experience of the Dao that the makers of China's gardens tried to make possible within the confines of their garden walls.

Tang Poets: the Tradition Grows Further

The Tang dynasty (618–906) saw the continuation both of great country estates and of the simple cottages of less well-heeled gentlemen. The former reached their highest point in a country villa on the banks of the Wang River not far from the capital at Chang'an. This *ferme ornée* belonged to Wang Wei (699–761), who was not only a scholar, poet and musician, but one

横屋漏

茅葺燥

濕藤萼

蘙蕫童

顏坐夫

手一巻

知是黃

庭内於

One of ten 'Scenes of a thatched lodge', attributed to Lu Hong, eighth century (detail). The poet Bai Juyi's mountain cottage may have been something like this, one of the earliest paintings of such a subject that has come down to us.

of China's most influential painters. In fact future generations would consider him one of the two great originators of landscape painting, and later we will see how the scroll he made of his estate affected the ideals of both painters and gardeners right up to this century.

Meanwhile, politics remained as tricky as ever, and exiled or temporarily demoted scholars seem often to have spent their time out of office building cottage retreats. The great poet Bai Juyi (772–846), of whom Arthur Waley has written so sympathetically,[17] has left a record of the 'grass hut' he built in 817 high among the huge old pines and waterfalls of Mount Lu (Lu Shan). It contained just two rooms and four windows and was furnished with four simple wooden beds, two lacquered lutes, and two or three volumes of the Confucian classics and Buddhist scriptures. In front there was an open area some hundred feet square, with a flat terrace in the centre and a square pool below surrounded by mountain bamboos and wild grasses. In the pond the poet planted white lotus and bred white fish for his table, and through the thick green creepers beneath the pine trees he made a winding path of white stones. On the day when he first moved in, he sat from seven o'clock in the morning until five in the afternoon, enraptured with the peace and beauty of the place. It was the most relaxing place he had ever been in, and he vowed to return and live there when all his official days were over (although as things turned out, he only managed to visit it once more, ending his days in quite a different garden). Like many other

95

poets after him, Bai Juyi was a true gardener, who in every new house seldom failed to plant a few bamboos or dig a little pond, however simple – or elegant – the circumstances of his life at the time.

Some two hundred years later another poet-politician, Su Dongpo (Su Shi, 1037–1101), also carried on the tradition of scholars' gardens. Like Bai Juyi before him Su Dongpo was made governor of Hangzhou and Suzhou, and he too fell in and out of favour in politics. At one time his official salary was totally cut off, and like Tao Qian he found himself living off such vegetables and grains as he could grow himself on a few acres of waste ground. Though the world was hard, the poet struggled through, happily calling himself the 'Layman of the Eastern Slope' (*Dongpo jushi*). He even found time to build two pavilions above and below his little house. Luckily, his next-door neighbour had a bamboo grove, where in summer the poet would lie and marvel at the brilliant green of the new shoots breaking through the earth, or the way the dew glittered on the rice stalks of his herds in moonlight. To Su Dongpo belong a number of poems very close in spirit to those of Tao Qian, such as this one, that expresses to perfection the ideal of the recluse's pavilion:

'How can you pass such days of peace and calm
While human life is so beset with ills?'
Last night I slept by the breezy northern window –
This morn the crisp air fills the western hills.[18]

Another great scholar-gardener of Su's time was Sima Guang (1019–1086), leader of a conservative faction in politics which suffered several serious reversals of fortune. In the tradition of Confucius, when this one-time Chief Minister realized that he could have no further useful effect on official life, he retired from the fray and devoted himself to writing a history for future generations. Indeed, he became one of China's most famous historians, and in the process added another name to the list of much-loved scholars' retreats: the Du Le Yuan, or Garden of Solitary Enjoyment, where he went to relax from his labours. Sima Guang's garden, although some distance from his house, was not situated among lofty peaks or in farmland beyond the city, but was right inside the walls of Luoyang. In fact it was an urban garden, but nevertheless one in which the owner tried to create an impression of country simplicity. It was not meant to be a

The Garden of Solitary Enjoyment (Du Le Yuan) of Sima Guang, by Qiu Ying, sixteenth century (detail). Ming dynasty painting of the celebrated Song scholar's garden. It shows the historian in all the various parts of the garden he described in his record of it. The painter has aimed to express not only the refinement and simplicity of the place, but its owner's delight in such ordinary pleasures as fishing, or resting from the heat in a bamboo grove.

great pleasure garden for elaborate entertainments, but on the other hand it represented something more than the extreme frugality of a Confucian sage: after all, he says modestly, 'the enjoyment of a sage could not be matched by a foolish person.' As a true scholar Sima Guang's real extravagance lay in the library of more than five thousand volumes which he collected in a hall in the garden. But he also spent many delightful hours designing the simple effects of his various pavilions and streams. His fish pond was dredged out in the shape of five small bays, like a tiger's paw. In its centre was an island on which he planted bamboos in a circle to look 'like a jade ring'; then he tied all the tips together to make a leafy tent and called it his Fisherman's Hut. From a little room built on a terrace Sima Guang could look out beyond his garden walls and see part of the city and distant mountains. For the rest, he can speak for himself – in the third person, which was his way:

> He usually spent a lot of time reading in the hall. He took the sages as his teachers and the many virtuous men (of antiquity) as his friends, and he got an insight into the origins of benevolence and righteousness, and investigated the ins and outs of the Rites and of Music. . . . The principles of things gathered before his eyes. If his resolve was weary and his body exhausted, he took a rod and caught fish, he held up his sleeves and picked herbs, made a breach in the canal and watered the flowers, took up an axe and cut down bamboos, washed his hands in the water to cool himself down, and, near the highest spot, let his eyes wander to and fro wherever he pleased. Occasionally, when a bright moon came round and a clear wind arrived, he walked without any restrictions. His eyes, his lungs, his feelings were all his very own. . . . What enjoyment could replace this? Because of this he called the garden the Park of Solitary Enjoyment.[19]

The enjoyment of such a garden was both contemplative and active, scholarly and sensual, and none of its delights presupposed an enormous fortune. Indeed, part of the Chinese genius for 'living harmoniously' comes from a profound desire to make the most of everyday existence, and from that particular attachment to the cultivated use of leisure which found its ideal environment in gardens.

Perhaps the best depiction of this feeling is in a little book called *Six Chapters of a Floating Life* written at the end of the eighteenth century. It is the record of a life spent almost entirely in

Engraved stone from the entrance porch,
Cang Lang Ting (Blue Waves Pavilion),
Suzhou, 1044, showing the garden's
eleventh-century layout.

Gallery around deep pool, Cang Lang Ting, Suzhou, 1975. The layout today is almost identical to that shown in the engraving opposite from the eleventh century. About one third of the garden is taken up with a false mountain on which stands the Cang Lang pavilion itself.

cultivated poverty, in which it was considered far more valuable to have a pot of chrysanthemums on the table than meat for dinner. The author, Shen Fu (1763–after 1809), was an impractical romantic from Suzhou whose efforts to earn a living were dogged by failure. His life was often tragic, involving the early death of his wife Yun with whom he was passionately in love, the loss of his parents' affection through bigoted misunderstandings, and the death of his father before they were reconciled. Later, a younger brother also cheated him out of his inheritance. Often Shen Fu found himself living off the charity of friends. Yet his book is full of irrepressible laughter and tenderness, describing people who never ceased to find delight in their own ingenuity and in the most simple joys of life. Gardens play a great part in it, often providing the background to the author's happiest moments. Once, for instance, Shen Fu, with the permission of his mother, spends a summer living with his wife in an open pavilion next to the gardens of the Cang Lang Ting (Blue Waves Pavilion), which still exist today:

> Beyond the eaves of the pavilion an old tree raised its gnarled trunk; its branches throwing a dense shade across the window, dyeing our faces green.... We spent the long, hot summer days together; doing nothing but reading, discussing the classics, enjoying the moonlight or idly admiring the flowers.... In all the world, we thought, no life could be happier than this.[20]

An earlier writer, Chen Haozi, who called himself the Flower Hermit, even went so far as to say that 'if a home has not a garden and an old tree, I see not whence the everyday joys of life are to come'.[21]

This was a commonly expressed feeling: the simple pleasures of a garden, away from the cares of men, could seldom be equalled. Impoverishment was no disgrace in a scholar, because his finely developed sensibility and literary acquaintance with all the great men of the past were considered infinitely more valuable than the mere accumulation of wealth. In fact writing was considered the essence of civilization; whereas our word 'civilization' comes from the Latin root meaning 'citizen' and 'city', the Chinese equivalent, *wenhua*, means 'the transforming power of writing'.[22] Public space in China has been traditionally not a physical but a mental arena, and even today different opinions find acceptable expression not in some equivalent of Speaker's Corner in London's Hyde Park but in handwritten wall posters. Such mental activity has always been extended over time, so that men like our garden makers Su Dongpo or Sima Guang did not feel themselves failures when out of political favour in their own lifetimes, because they were part of a much wider debate that took in thinkers of the future as well as those of the past whose work they read. Indeed, the scholar's retreat did not only represent the gentleman's appreciation of nature but can also be seen as a symbol of this 'mental space', and thus of the cultivated aspirations of all educated Chinese.

The Elaborate Tradition Continues

Meanwhile, however, rich and great families also continued to build gardens to show off their wealth in the tradition of Yuan Guanghan, and vied with each other in the number and expense of their pavilions and decorations. At the beginning of the ninth century Bai Juyi wrote of gardens of this kind that had been abandoned by great families whose interests took them elsewhere:

> By woods and waters, whose houses are these
> With high gates and wide-stretching lands?
> From their blue gables gilded fishes hang;
> By their red pillars carven coursers run.
> Their spring arbours, warm with caged mist;
> Their autumn yards with locked moonlight cold.
> To the stem of the pine tree amber beads cling;
> The bamboo branches ooze ruby-drops.
> Of lakes and terrace who may the masters be?
> High officers, Councillors-of-State.
> All their lives they have never come to see,
> But know their houses only from the bailiff's map![23]

Such gardens were a long way in feeling and purpose from the scholar's hut, yet the impact of this simpler image was so powerful that it began to be absorbed even into the Imperial parks and the aristocratic pleasure grounds of Bai's poem. Li Deyu (787–850), a contemporary of Bai's and a great aristocrat and politician, owned a great and famous villa called Ping Quan (the Level Spring) which was both an extensive park and a literary retreat. Here he grew not only many exotic plants brought from the south, but also medicinal herbs and trees, and millet and tea. Although Li was politically opposed to Bai Juyi, their personal tastes were in perfect accord. Both were great fanciers of strangely shaped stones, and are thought to have made popular the appreciation of single rocks in garden settings. Like Bai, Li Deyu loved nothing better than to wander over his estate, and wrote movingly of his happiness when, returning from exile in the south, he found his chrysanthemums and pines still waiting for him. He had his poems engraved on to some of the finest items in his stone collection, and like Bai left a record of the mountain scenery around his villa and of the many rare shrubs and trees he had coaxed to grow there.

It was, however, in the Song dynasty that the different strands of elaborate pleasure ground and scholars' retreat were finally woven into their characteristic form. This occurred in the famous city gardens of Luoyang, later to be described in a book by Li Gefei. Each of these city gardens (except one which was an annual flower market rather than a proper garden) had many fine halls and pavilions beautifully set among lush plantings of shrubs and trees. Li Gefei praised some for the exceptional beauty of their sites, others for the size of their aged pines or the lush shadows of their bamboo groves. In particular he noted those gardens which had belonged to famous scholars: one of Bai Juyi's gardens, for instance, was still extant with its layout unchanged, although the original pavilions had long since fallen down. Sima Guang's old garden also still existed, but Li Gefei was disappointed in it. 'It was rather small', he wrote, 'and could not be compared with the others. . . . The fishing cottage and the herb garden were but constructions of bamboo branches and creeping grasses done by Sima Guang himself.'[24] He concluded indeed that the garden's popularity must have had more to do with the account the historian wrote of it, than with any intrinsic qualities of its own. For Li Gefei, it seems, do-it-yourself simplicity did not stand up to the elegant, supervised simplicity of more elaborate gardens, in which the ideals of a scholar's retreat had been incorporated into the grand tradition of the wealthy official's garden. The gardens in Luoyang he most admired had streams channelled into circular moats, stone lotus flowers gushing water, and high towers for viewing the hills. One garden even contained a building so complex it was like a labyrinth, while the one with the most beautiful scenery of all he praised for the density of its layout as well as the richness of its owner's imagination.

Most of these famous Luoyang gardens seem to have been open to visitors, even if the owner was not there himself and not personally acquainted with the visitor. Indeed, one garden is thought to be remarkable because it has been abandoned to caretakers – like the villas of Bai Juyi's poem – and is 'locked and closed to society'.[25] In later novels and stories it is quite common to find descriptions of outings to famous private gardens in the city or the suburbs,[26] and for many people in China such trips seem to have been genuine substitutes for excursions into nature. So, despite the rustic huts which in Chinese paintings are the focal point of countless magnificent landscapes, it was always just as common for a scholar to retreat to a garden in the city or the suburbs as it was for him to take to the wilds.

It was, of course, far more convenient that way, and enabled the gentleman to enjoy his solitude in the company of friends – with several jars of wine – instead of wasting it on the wilderness. Nevertheless, the illusion of being far away and surrounded by nature, which one would expect to work against the tradition of elaboration, paradoxically came to augment it. For the Chinese scholar–gardener still aimed, as his Emperor did, to re-create the effect of the *totality* of nature in his garden. And to do this he became increasingly involved in creating an infinite number of different experiences in a small space: so evolved those intricate labyrinthine gardens which to us seem so remarkably artificial, and yet to the Chinese are still 'natural'. This is partly because, regardless of how fancy they may look to us, to the Chinese they still recall, through allusion and traditions, the simple retreats of great poets and scholars. But it is also partly because their visual conventions, developed first by China's landscape painters, were seen as distillations of natural forms. We shall see how these conventions developed in the next chapter.

THE PAINTER'S EYE

In China landscape painting and gardening are so intertwined in their development that it is hard to appreciate the one without knowing something of the other. Both arts developed together, painters providing several of the conventions through which the Chinese looked at their gardens, and gardens in turn giving back these conventions to painters. Together they created an 'eye' for enjoying precipitous mountains, and highlighted the moods of changing seasons. Paintings also developed the two great garden themes: the lone scholar's hut, and the luxurious, manmade landscape with all its decorative features – including beautiful women. We see gardens differently once we know painting.

Early Landscapes

The first traces of landscape in Chinese art are the conical hills and willowy trees that make their appearance in the backgrounds of hunting scenes and in representations of the Immortals. They are grouped and dotted around the figures, separating them into different scenes to make what critics call 'space cells' in the pictures.

In the eighth-century painting of *Emperor Ming Huang's Journey to Shu* (opposite), mountains and trees arranged like this take up at least two thirds of the space, but it is still the travellers and not the landscape that form the subject of the painting. Historically the Emperor was fleeing from an uprising, and the journey through the mountains must have been arduous and terrifying. Perhaps the artist meant to convey this by the craggy precipices in his painting, but his glowing enamel colours turn the whole scene into an exquisite fantasy. The tiny figures seem to be winding through one of those bibelots designed by Fabergé for the Russian court. In the foreground pines and blossoming plum trees sit precisely on flat slabs of rock. Above them, sharp clusters of fractured peaks rise up in needle points, or are cantilevered incredibly over space. Stiff white clouds separate the peaks from each other, while a pure and shadowless light illuminates the whole scene with unnatural clarity.

This painting is an early masterpiece in what is known as the 'blue-and-green' style. Although it certainly does not suggest the kind of garden design Europeans would call 'natural', its courtly elegance may have influenced early attempts to recreate the effect of great mountains in Chinese palace gardens. At least two early gardens are said to have contained polychrome rocks. One of these belonged to a prince who briefly ruled the state of Qi at the end of the fifth century AD. He was famous for his extravagance, and although there is no way of proving this, it is possible that his taste for coloured rocks was encouraged more by early blue-and-green paintings than by mountains in the wild.

Later, in the sixteenth century, as we shall see, the painter Qiu Ying brought the blue-and-green style to a new height of elegance and grace, giving it an even greater mixture of fantasy and precision, but his use of it was consciously archaic. Already by the eighth century painting in outlines and washes of monochrome ink had begun to develop, particularly in the hands of the great poet-painter Wang Wei (699–761).

The First Master of Monochrome

There is a famous phrase in the Taoist classic *Dao De Jing*, which goes: 'The five colours dazzle the eye, The five tastes confuse the tongue'.[1] Buddhists, with their doctrine of Nirvana, felt the same way, and to Wang Wei, who was a devout Buddhist, dazzling the eye was a sure way to confuse the soul. Though he may not actually have invented monochrome, he certainly developed it further than anyone before him. Later painters were deeply affected by the luminous subtlety of his paintings of mountains under snow, from which all colour is excluded. As monochrome developed it seems likely that it also affected the way people looked at gardens. To eyes accustomed to landscapes expressed in washes of grey ink, even when they depicted scenes of high summer, a blazing vision of variegated flowers might well have begun to seem crude and aggressive. In the great Suzhou gardens that still exist today, colour is used quite sparingly. Even when all the shrubs and trees are in bloom, one's attention is consciously focused not so much on colour as on light and movement, on the play of sun and shadow among the grey rocks, the dancing patterns of water-reflected sunlight, the shifting outlines of leaves on a white wall. Green, grey and white are the predominant tones of these gardens, with highlights of seasonal colour rather than a continuous polychromatic effect.

Instead of seeing mountains only as decorative shapes, Wang Wei's monochrome forced the painter to examine their structure. To bring out the geological formation of hills under snow, he is credited with introducing a new style known as *po mo*, or 'broken ink', in which the rock outlines were no longer flatly coloured in but built up out of different ink washes. With this

Wang Chuan scroll (detail), copy after Wang Wei (699–761). Each 'space cell' of the long scroll has some manmade object as its focus, and each corresponds to one of the verses in which Wang Wei described his villa: 'Dwelling here,' he wrote, 'why should I ever leave home?'

ABOVE   *Tiny courtyard off a larger enclosure, Zhuo Zheng Yuan, Suzhou. The garden is divided into innumerable different 'space cells' enclosed by walls, planting, hills, buildings. Some are tiny like this one, others so arranged with rocks and trees that the visitor is unaware of the boundaries.*

ABOVE, RIGHT   *Entrance courtyard, Yi Yuan, Suzhou. White walls rise behind rocks and trees like mist, or the empty silk of a landscape painting.*

method went a brushstroke called the *cun,* a term which is untranslatable but means something like 'wrinkles on the mountain face'. These two techniques enabled the artist to express the effects of age and erosion on his hills. From then on people noticed the way mountains were formed, and the painting of mountains became more and more expressive. It seems likely that this also began to affect the choice and arrangement of garden rockeries.

The most famous of all Wang Wei's works is a portrait of the country estate he owned on the Wang River, or Wang Chuan, some thirty miles from the capital. For anyone interested in Chinese gardens it is a key painting: Wang Wei was not only the master to whom all later scholar-painters would refer, but his country retreat became (in Wan-go Weng's words) 'the summation of the private domains of all scholars. Every scholar would like to see something of the Wang Chuan around him.'[2]

Unfortunately, none of Wang Wei's original works survive, but a later copy of the Wang Chuan scroll in the British Museum shows that the estate was made up of different pavilions and dwelling houses tucked into a soft but dramatic landscape of rounded peaks and gentle river banks. Although it is the most famous painting by the first master of monochrome, the original seems to have been in the old blue-and-green style, and Wang Wei is still clearly drawing on past landscape conventions. But whereas the landscape of *Emperor Ming Huang's Journey to Shu* is treated as a piece of jewellery, Wang Wei is making a personal record of a place he loved intimately. He had walked over the hills he paints, and composed poems in the different places he highlights. In his painting he is beginning to express the sort of personal relationship with the landscape which had already been expressed by poets. His closeness to the mountains and woods is shown in the new techniques he uses to reveal their character. It is not only that his hills rise to less incredible peaks than those in the *Emperor Ming Huang's Journey,* but that they now seem to grow out of the ground, rising massively, as if pushed from below. And the dividing mist lies realistically along the valleys, dissolving and reforming over land and water as he must have seen it many times on his estate. He is clearly reaching for a new way to express the 'natural'.

As well as his other innovations, Wang Wei is also sometimes credited with inventing a new format: the scroll painting which one unwinds from right to left. Each space is separated, but linked to the next by water and mist. As the scroll is unwound the landscape passes by so that a temporal dimension is added to the static painted forms. Like music or a film, and unlike Western painting, the handscroll and the garden it depicts are experienced over time. The same effect was later sought in gardens themselves: the garden, its space cells divided by white walls simulating mist, became a three-dimensional walk through a landscape scroll.

Wang Wei's stature in Chinese art history is due to his exemplary character as much as his technical innovations. For later ages he was the ultimate gentleman – brilliant, cultivated, unworldly and a model of filial piety – and the Wang Chuan villa, like all his other works, was seen as an expression of this. The idea needs stressing: it was neither his art alone nor his character alone which was celebrated, but the conjunction of the two, each as a sign of the other. The quality of the painting, the garden, the music, the poetry, the calligraphy – for later scholars all these reflected the quality of the man. Later, as we shall see, this idea will play a vital role in the development of painting, and come to affect too the way people felt about gardens.

The Vital Spirit

For Wang Wei's contemporaries his paintings were so realistic they were almost like magical incantations calling up a synaesthetic experience. Looking at them they could actually hear the splash of water, feel the stillness of space, catch the fresh scent of pine trees on the breeze. In short the paintings seemed animated with a kind of life of their own, and the feelings they conjured in the viewer were truly those he experienced in real nature.

This was in accordance with the first of six famous principles for judging the quality of painting which had been codified in a sixth century essay by Xie He.[3] According to this, the first condition of 'good painting' is *qi yun sheng dong,* translated by Alexander Soper as 'animation through spirit consonance'.[4] According to him Xie He's criterion can mean two things: that the *qi* or 'vital spirit' of every different thing in the painting must 'vibrate' with the *qi* of all the other elements; and that the *qi* of the painted forms must also respond to the real forms as they exist outside the painting. The complete principle, therefore, is one of inner consistency, by which themes echo each other and set up 'harmonious vibrations' among themselves, but also one of mystical realism, by which the artist magically catches the living spirit of nature itself. Clearly, the idea owes a lot to Taoism, whose sages aim to become one with the rhythms of the universe.

In Wang Wei's time this first principle was still closely bound up with magic. A contemporary, Wu Daozi, was said to have painted such life-like landscapes that he actually stepped into one of them and disappeared. However, critics looking back at the works of earlier masters began to find themselves in difficulties: what was manifestly 'real' to one generation began to look strangely stiff to the next. At the end of the Tang dynasty the critic Zhang Yanyuan even found himself arguing that it was not the old masters' fault if their pictures looked odd, for it was nature, not art, that had changed.[5] Later, however, he found it possible to explain that the 'vital spirit' did not necessarily reside in formal likeness; many old painters indeed had managed to transmit the life force, even though their trees and hills now seemed unrealistic, while many contemporary painters caught the likeness of things without communicating their essential vitality.

In this way *qi yun sheng dong* proved to be such a suggestive and usefully ambiguous phrase that it could be almost endlessly adapted to suit developing theories. The discussion is taken a stage further around 960, by Jing Hao, who first underlines the spirit that lies behind all successful representation and then classifies this into four main types; the 'divine', the 'sublime' (or 'mysterious'), the 'marvellous' and the 'skilful'. In the class of the divine he says 'There appears to be no trace of human effort, hands spontaneously produce natural form'.[6] Here the

RIGHT ABOVE   *Taoist monks above North Peak Monastery, Hua Shan. The z-bends of the ridge, rising out of misty chasms below like the back of a dragon, recall the craggy peaks of the landscape painting, far right. (Photographed in the 1930s by Hedda Morrison)*

RIGHT BELOW *A dramatic rocky ridge on the sacred mountain Hua Shan, in Shaanxi.*

FAR RIGHT   *'Buddhist Temple amid clearing mountain peaks', by Li Cheng, tenth century. Northern Song landscape painting at its best. A vast precipice looms over a Buddhist temple; trees with crab-claw branches are perched way out on crags which echo their shape. Below, tiny peasants approach an inn on the water. The architecture nestles into the scene, and with its upturned eaves also takes up the spiky theme of the trees.*

PAGES 108–9   *Detail from a scroll painting by Qiu Ying, c. 1510–51,* Festival of the Peaches of Longevity. *Youthful Immortals in the gardens of Xi Wang Mu, the Queen Mother of the West, pick the peaches of longevity which ripen every three thousand years. For another garden painting by Qiu Ying see pages 96–7 and pages 123–4 for a discussion of his effect on landscape gardening.*

'Eight Riders in Spring' (detail), anon., tenth century. Aristocratic horsemen ride through an imperial park, suggested here by the elegant railings and extraordinary rock behind them.

artist is so perfectly in accord with the spirit of his subject that his own act of creation is identical with that which produced the original forms in nature.

For Jing Hao the viewer's attitude was equally important, for emotional response was directly transferable back and forth between viewer and painting: the quality of the one meeting and resonating with the quality of the other. Woe betide a man, therefore, who created or contemplated a less than merely 'skilful' work, one which was 'highly exaggerated . . . owing to the poverty of inner reality and to the excess of outward form'. The spirit resonance of such a person, his own *qi yun*, was seriously deficient, or at least in need of servicing.

All these reverberations from the *qi yun sheng dong*, or first principle of painting, suggest some fascinating comparisons with Chinese gardens and how they might have been perceived as the theory of painting developed. First of all, successful gardens would have had a double 'vital spirit' – that supplied by nature, in the living rocks and trees, and that 'harmonious vibration' which the garden designer achieved by careful selection and composition of garden elements. Like a painter, the gardener's aim was more than to reproduce nature on an intimate scale; for although the garden manual *Yuan Ye* says 'Try to make your mountains resemble real mountains', another recommendation immediately qualifies it: 'Follow nature's plan to a certain extent, but do not forget that it is to be executed by human hands. Select the peculiar and seize upon what is good. Those who have the right interest will understand the matter.' Like the painter, the garden maker must seize on the *best* bits, and through them grasp the essential. But he must avoid the particular, for his mountains are to stand for all mountains, expressing the deeper reality, not their individual likenesses. In other words he is to be an interpreter, not a copyist – or as a Chinese critic once expressed it, a novelist, not a photographer.

The Northern Song Painters

This ideal was achieved in its most heroic form during the first 'Great Age' of Chinese landscape painting in the tenth century. In the few paintings that survive from this time, huge tumultuous mountainscapes dominate all other elements. Although Wang Wei's hills are large, they are still psychologically a kind of frame for the more important theme – the activities of man. However, in northern Song paintings mountains take over centre stage, often literally covering three quarters of the scroll, which is turned vertically to allow for their massive peaks. The northern landscape painters do not exaggerate in their representations, for such incredible mountains do exist in China. In photographs of the five great holy mountains, cliffs tower above great misty chasms, and tiny buildings sometimes cling in the crevices or balance precariously on knife-edged peaks. Although pictures (from the 1920s) show Taoist monks with tranquil expressions, their depiction of uninterpreted nature is profoundly disturbing. Are the travellers and peasants protected by these vast manifestations of natural power, or threatened by them? Are the peaks meant to defend the temple or intimidate the priests? In these photographs the subject matter is exhilarating, yet terrifying. In the paintings, however, something quite different happens. The raw forms of nature are transformed into a harmonious, symbolic vision of the Dao itself. The workings of the Dao could be understood through the extraordinary coherence of the painted forms: entering into such a painting thus became a spiritual, almost religious, experience, for the viewer as well as the painter.

All this helps to explain those extraordinary monolithic rocks in Chinese gardens, which will be discussed more fully in a later chapter, 'Rocks and Water'. We have seen that even a single rock in a garden may represent a complete range of mountains and concentrate their essence. In the tenth century one such stone appears in a picture called 'Eight Riders in Spring'. The

foreground shows aristocratic horsemen riding through an imperial park. Behind them, the top half of the picture is completely taken over by a fine balustrade, three trees and a great grey garden rock. The stone billows out of the earth – its stem stringy and muscular, its top soft like the vaporous bulges of some enormous fungus. It is hard to see how anyone could have seen a rock like this as a representation of even such fantastic mountains as can be found in China. To have this vision one needs not only a 'Drink Me' bottle like Alice, to shrink down to the size of a flea, but also to have seen mountains through the eye of Chinese painters. In a painting attributed to Fan Kuan, called *Scholars' Pavilions in the Cloudy Mountains by a Stream*, the artist builds up the mountain from bulbous silhouettes which rear up out of misty space as if they were quite unconnected with each other. If you half close your eyes, and imagine a garden rock with its crevices half filled with mist, it becomes easy to see what the rock-fanciers drew from painting. Tong Jun is explicit about this connection: 'The question of reality does not really bother [the visitor], as soon as he ceases to be in the *garden,* and begins to live in the *painting*.'[7]

Chinese artists were not slow to exploit the enjoyable ambiguities of such a situation – painting pictures of gardens which were themselves illustrations of paintings. When around 1080 Li Gonglin immortalized a number of famous scholars in the Western Garden of a well-known art collector, he deliberately played on the relationship between the 'real' hills and peaks,

ABOVE, LEFT *Scholars' pavilions in the cloudy mountains by a stream, by Fan Kuan (c. 990–1030). Bulbous silhouettes loom out of misty valleys.*

ABOVE, RIGHT *Single Taihu rock on a marble pedestal, Yu Hua Yuan (back garden, Imperial City), Peking. If you half shut your eyes and compare this rock with the painted mountain next to it, it becomes easier to see how painting influenced the 'eye' of rock collectors.*

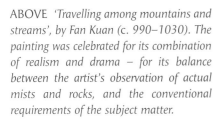

ABOVE *'Travelling among mountains and streams', by Fan Kuan (c. 990–1030). The painting was celebrated for its combination of realism and drama – for its balance between the artist's observation of actual mists and rocks, and the conventional requirements of the subject matter.*

RIGHT *'A poetical meeting in the Western Garden' in 1080, by Zhao Mengfu after Li Gonglin. The painter and rock-connoisseur Mi Fu added an inscription to this picture identifying the famous guests. Best known is the poet Su Dongpo, here inscribing a boulder in the middle of the garden – a miniature version of the great peak in the painting seen left. Note also the way the tortured pine, a convention of painters, has come into the garden which, despite its formal gateway, high wall and edged canal, is made to look as 'natural' as possible.*

which seem to be visible beyond the garden, and the 'false' mountains which are scattered about the garden itself. He has most fun with the fractured peaks he paints in the background; one is a hazy outline, another sticks out of the hillside like an enormous tooth. Are they both outside, or both inside the garden? Is one a representation of the other, or are they in fact both 'wild'? Or for that matter, are both of them 'false'?

This particular garden seems to have been situated in the kind of area the *Yuan Ye* finds ideal: among the trees in the mountains. Such places, suggests the manual, 'offer wonderful scenery in themselves and do not need to be worked on with human hands'. However, it seems that by the early twelfth century, garden conventions were already growing stronger than nature: false mountains are necessary not only in city gardens but in ideal sites too. The owner of this Western Garden clearly feels the 'wild' hills surrounding his estate are inadequate in themselves. Perhaps their effect was a little too soft, too diffuse, too distant. At any rate the painter shows Su Dongpo, the great Song poet, inscribing a verse on an enormous, vertical slab-rock, which is a focus for the whole garden. It does not require much effort of the imagination to see its rectangular shape and weedy fissures in terms of one of those precipitous mountain cliffs of northern Song landscape painting.

Mood Painting

Perhaps the most remarkable mountain-painter among the Song artists was Guo Xi (died *c.* 1090). A great deal of his discourse on painting is taken up with the painting of mountains, which to him were 'store houses of the world, in which are found the treasures of Heaven and Earth.' For Guo Xi any artist worth discussing should identify himself completely with his subject, his spirit resonating with the mood of the mountain, the mountain with him. Of course, this is the *qi yun* again, but Guo Xi takes it so far that we seem to be confronted with something close to Western pantheism. Human feelings are projected onto the physical forms of nature in a way that suggests our nineteenth-century 'pathetic fallacy', and the poetry of Coleridge and Wordsworth. Nature becomes man's psychological state writ large and in vegetable form, and the whole weight of creative responsibility begin to shift from technique, or even natural ability, to the artist's character and mood. To paint great works, an artist must now 'nourish gentleness and cheerfulness in his heart; his ideas must be joyful and harmonious; only that way will his spirit resonate truly with the spirit of his subject, and the various aspects of man's gladness and sorrow . . . will appear naturally in his mind and be spontaneously brought out by his brush'.[8] Nature enters into a two-way relationship with man's deepest feelings; the outer world changes according to his emotions, and his emotions follow the moods of nature like a barometer.

The transference of emotion between nature, art and man, which previously had always been an important principle, now became *the* principle. The result was mood painting, highly charged with emotion, often striving for expressionistic effects, and often, to the untrained eye, wildly exaggerated. Guo Xi did not fear the consequences of elaborated detail for he felt that the human eye 'had the power to grasp all the details in the foreground . . . and encompass a wide view', in painting as in life. Besides, what mattered was always to capture the *vital spirit and movement of life.* For Guo Xi 'realism' meant being true to an emotion:

> The spring mountain is wrapped in an unbroken stretch of dreamy haze and mist, and the men are
> joyful; the summer mountain is rich with shady foliage, and the men are peaceful; the autumn
> mountain is serene and calm, with leaves falling, and men are solemn. . . . The sight of such pictured
> mountains arouses in man exactly corresponding moods.[9]

'Early Spring' (detail), dated 1072, by Guo Xi. 'The spring mountain is wrapped in an unknown stretch of dreamy haze and mist,' wrote the painter, but here the whole landscape seems to be writhing in turmoil: grotesque faces leer out of crevices. Close inspection, however, may reveal tiny figures scattered around the lower slopes and a peaceful temple and village above the waterfall at centre right.

Interestingly, this is the technique used by Ji Cheng, the seventeenth-century author of the *Yuan Ye*, when he describes different kinds of gardens and sites. In each case he paints a word-picture so as to create in the potential garden maker not so much a precise image, as a particular emotion. So, for a garden in a noisy town, he writes that one should choose a secluded spot offering an extensive view: one must be able to see remote mountains standing in a 'curving row, like a screen. . . . There are fairies everywhere. . . . Roll up the bamboo blind and let the swallows fly in with the wind. The petals of the flowers hover like snowfalls. . . . Let your feelings dwell among hills and valleys; there you may feel removed from all the unrest of this world. In your fancy you enter a painting.'[10]

Although Ji Cheng does suggest some positive elements to be included in such a garden, the descriptions become more and more impressionistic. They are hardly prescriptions for gardens, but rather attempts to put the designer in the right frame of mind. When he comes to actually build the garden he will thus become so much a part of the landscape he aims to create that the physical forms will seem to have appeared of their own volition.

Lyric Spontaneity

When the Emperor Huizong came to the throne in the northern Song the taste for tumultuous mountainscapes was replaced by paintings of more intimate scenes such as birds in garden settings (as we have already seen in the chapter on imperial gardens). After his fall, when the court moved south to Hangzhou, the Painting Academy that was re-established there continued to develop these softer themes.

Under the influence of the beautiful southern capital, and the philosophy of Chan (Zen) Buddhism which flourished in the monasteries around the Western Lake, two new categories of painting are added to those we have already mentioned: the 'lyric' and the 'spontaneous'. Landscapes become ever more personal as painters accentuate the atmospheric qualities of mist and disappearing space – until empty space indeed threatens, like a modern abstraction, to eat up the whole painting. Objects are painted with such deft economy of brushstroke, such terse and spontaneous flicks of the educated brush, that after months of preparing for the right moment they are finished in a second.

Several painters of the Southern Song Academy continue the theme of the scholar in nature contemplating a 'real' waterfall or a 'wild' pine tree. But he is so much part of the scene that it is hard to tell whether he is part of nature, or nature part of him. A painting by Ma Lin, *Listening to the Wind in the Pine Trees,* shows this moment of communion when the scholar and nature have happily arrived at a state of perfect empathy. Scholars sometimes achieved such heights with the aid of a little alcohol. Here, however, the gentleman seems a little tense and self-conscious, listening to the sounds of water and wind, while he steals a sidelong glance at his boy attendant. The elastic zig-zag of the pine tree, set off against the smooth serenity of the distant mountains, takes up the contradictions in his mind. The subject of the painting has become the inside of the painter's mind, his own 'vital spirit' projected on to the world in which he sits.[11]

The painters of the next dynasty, the Yuan (1280–1368) took such moods one step further. Cutting away the sophisticated ambiguities of such painters as Ma Lin, whose work they regarded as too slick and clever, they produced a spare, aristocratic manner, an esoteric reticent style that is as important for what it leaves out as for what it includes. From now on, in autobiographical landscapes, there may not be a figure at all: mountains and waters themselves stand for the character of the man.

The idea that an artist's character was directly reflected in his works was hardly a new one. It had been inherent in the theory of Chinese painting from the beginning. But in the eleventh century this had been further developed by Su Dongpo (the poet we have already met in the Western Garden and working his own land as the 'Layman of the Eastern Slope'), into an ideal of the amateur in art. A friend of Su Dongpo who called himself the 'Hermit of the Western

'Listening to the wind in the pines', by Ma Lin, 1246. The mood of the artist projected on to nature. The sound of pine needles shifting in the wind is a favourite image in poetry and painting.

Shi Zi Lin (Lion Grove) garden, Suzhou, by Ni Zan (fourteenth century). The sparse, deliberately sketchy brush strokes bring out the conscious simplicity of the garden, built by a monk. The celebrated rock-work, arranged to look like irregular stone cliffs, forms a private enclosure between the main hall and the small study behind it. The present-day Shi Zi Lin is very different from this.

Lake', perfectly expressed this idea. He was a 'genius who never passed any examinations, and never sold any pictures'. 'I write,' he said, 'in order to express my heart, I paint in order to satisfy my mind: that is all.'[12] In fact Su Dongpo and many of his brilliant circle were not hermits, as they liked to refer to themselves, but officials of high standing. However, the examinations by which they entered government service had not been designed to test anything so mundane as administrative ability. Thus, in a sense, they were amateurs in their official life just as much as they were in art. Literary prowess and knowledge of the classics were what counted, for such studies, it was confidently supposed, would develop the character of a man and enable him to make far-reaching decisions much better than would mere technical skill. People applied the same idea to gardening, and gardens, like paintings, were valued for the character of their creator as well as the intrinsic brilliance of their parts or conception. Because fame depended (in theory at least) on character, the Chinese could discuss this idea with the accent on reputation: 'If its owner enjoys lasting fame, his garden will stand through the ages, even though it may fall into decay'. It was not only the owner who could reflect such credit on a garden, however. If a famous man had been associated with the place, that in itself would often be sufficient recommendation. Such reverence for the haunts of the famous is perhaps worldwide, but in China the particular stress on character added a further dimension, in that it included the idea that a man of impeccable qualities would not sit about in an inferior garden.

Today, two of the most famous gardens that survive in Suzhou – the Shi Zi Lin (or Lion Grove) and the Zhuo Zheng Yuan (Artless Administrator's Garden) – still stand out among the others because of their association with famous painters, Ni Zan and Wen Zhengming respectively. Each made small paintings of the garden he frequented. It is also said that Ni Zan himself worked on the arrangement of the rocks in the Shi Zi Lin, but a glance at his scroll shows that it was very different in his day from its present appearance. It is tantalizing to speculate whether, if any of his rocks had survived, modern connoisseurs would have been able to discern his character from them. If so, they would have been unravelling one of the most fascinating personalities in Chinese art, for his life and work – which according to the time-honoured formula were symbolic of each other – exhibited all the paradox inherent in the scholar tradition.

Ni Zan was born to a family of well-to-do merchants just north-west of Suzhou, and had all the possessions a gentleman could desire – weather-worn rocks, old paintings, bronzes, jades, a large library, many fine pavilions and several houses and gardens. One of the bizarre activities which came to symbolize his work was his incessant washing. He was so fastidious that he would have all his garden seats scrubbed down after the departure of visitors (in case they were not quite clean). No one was allowed into his own garden pavilion, his decontaminated sanctum, unless Ni Zan was sure the person shared his own refined and pure sensibility. Every day he washed himself several times and this purification ritual obviously cleansed more than his body. It is quite evident in his remarkably spare, austere painting. Moreover his austerity also prepared him for his later years: for in 1356, when in his early fifties, Ni Zan gave up all worldly possessions and with his wife took to a houseboat in which he toured the lakes and streams of Jiangsu province. Every so often he would stop off at one of his only remaining possessions, a little cottage which he called, ironically, 'Snail Hut'. His lack of worldly ambition perfectly agreed with the amateur ideal in art, and his act of renunciation fulfilled equally the Taoist ideal of revolt against the follies of society and power.

In its style and subject matter, Ni Zan's painting was also characterized by high-minded renunciation. At first sight it may look crudely done, lacking in realism. This was an effect of which he was inordinately proud. When faulted for drawing bamboo as if it were 'hemp and rushes', he answered that his only aim was to 'set forth the untrammelled feelings in my breast'. When criticized for being unrealistic in another picture, painted while very drunk, he answered with artistic one-upmanship: 'Ah, but *total* lack of resemblance is hard to achieve; not everyone can manage it'.[13]

The point of painting in such a purposefully charmless manner was to outflank the philistines, to put them off the scent so they would never be able to decipher the subtlety and brilliance lying hidden in the banal, the informal. All the better if self-satisfied collectors and Emperors failed to understand the beauty of the 'insipid' and 'clumsy' – two categories now elevated to sit alongside the 'divine' and 'sublime'. Virtuosity should only be apparent to one or two other like-minded souls, for mass appreciation might contaminate this virtue.

Ni Zan's work conveys an undeniable coolness and stillness, and that much sought-after quality of strength without the appearance of strength. Subsequent scholars tried often to work in his style, but although he is one of the most copied painters critics agree that ultimately his works were inimitable because they corresponded so completely with his unique life.

In garden art one would have expected Ni Zan to have renewed the old Taoist sensibility, the taste for the plain and undecorated, if not the 'bland' and 'insipid'. Certainly nearly all remaining Chinese gardens have some part celebrating the ancient virtues of the simple and everyday, the modest and quiet, which may owe a debt to Ni Zan's paintings. And the empty stillness, the sense of deep timelessness which the old labyrinthine gardens still manage to convey, mirrors the emotional purity of Ni Zan's character.

The Scholar-Gentleman-Amateur

This is certainly the effect of another garden in Suzhou dating from the early Ming period, which was also partly designed and painted by a great master. Wen Zhengming (1470–1559) is known as one of the 'Four Great Masters of the Ming', just as Ni Zan is numbered among the 'Four

ABOVE 'Landscape', by Ni Zan, fourteenth century. A characteristically cool work, in which space is infinitely extended beyond sparse trees, a farmer's hut, crisp bamboos and distant islands.

RIGHT Begonia-flower door, Shi Zi Lin, Suzhou. Simple but fanciful and edged in black, the shape of this doorway highlights the appeal of the courtyard beyond.

LEFT *Tiny courtyard, Shi Zi Lin, Suzhou. When the available space is too small, the Chinese trick is to divide it further, creating secret corners. Framed holes, like this 'moon' door, focus the eye and lead it through the wall, suggesting more space beyond.*

RIGHT *'The bank of many fragrances', album leaf of the Zhuo Zheng Yuan (Artless Administrator's Garden), Suzhou, by Wen Zhengming, 1551. Poems in fine calligraphy accompany these little paintings of a famous garden. In this one the painter-poet describes his pleasure in the scented air: 'My high thoughts are already beyond this noisy world. Aloof I watch the bees flying up and down.'*

Great Masters' of the previous Yuan dynasty. Like his predecessors Wen Zhengming continued the ideal of the amateur in art, and helped to develop it into the Ming dynasty's *wenren hua* – the 'painting of the scholar-gentlemen-amateurs', also known as 'literati painting'. Under the Yuan, a Mongol dynasty, such amateurs had had good reason for rejecting official society since the government was run by foreign invaders. Under the Chinese but repressive Ming dynasty the 'scholar without office' became an even more familiar figure, and he was nearly always to be found in a garden setting. Wen Zhengming was such a man. He was most famous for his austerity, his cool intellectual restraint, his high ethical standards, and it is these qualities of character that he particularly reveals in his paintings of the Zhuo Zheng Yuan. Like the artist, the patron to whom this garden belonged had retired from an unhappy political career. The name of the garden commemorates his feelings, for it can be translated either as the 'Unsuccessful Politician's Garden', or as the 'Garden of the Artless (or Humble) Administrator' – referring to the remark of a third-century official who also aspired, or said he did, to a life of retirement, to the effect that gardening was the form of administration best suited to the artless. While Wen retired penniless, however, his patron, Wang Xianchen, had acquired a fortune, and he made his garden available to the artist. It was a wise move, for Wen Zhengming brought the garden and his patron lasting fame as a result. He seems to have loved the Zhuo Zheng Yuan, and to console himself for not owning it he recorded his reactions to its changing moods in two series of little album leaves. This was a very personal format, particularly favoured by the *wenren* painters

because it combined poetry, calligraphy and painting in equal proportions. The style is deceptively simple, in the best tradition of ordinariness and insipidness. As topographical description the little paintings are not of much use, but as an expression of what a scholar-painter valued in his garden they could hardly be bettered. In quite a real sense, this garden is a portrait of the artist.

The real Zhuo Zheng garden lay inside the city walls, but it is treated here as if it were a country villa. In fact it appears even more domestic and pastoral than Wang Wei's country villa. It has chickens and vegetables and farmhouses with fruit trees, and in the first picture, so the accompanying poem tells us, the scholar and his servant are discussing the ploughing. Above their heads the outlines of the city walls hover over the garden roofs. In the real garden the castellations rising and falling above the white-painted wall would have become symbols for the mist and mountains of the wild. In this first painting they are more an initial reminder that the city streets have been shut out, that all this peace and rural charm is a manmade haven in the midst of town. However, this is the only time the artist makes any reference to the garden's boundary or location. Other pictures show walls inside the garden, with trees in courtyards on the other side, but never again an exterior boundary. Indeed, in the second leaf the distant hills beyond the city gate have become part of Wen Zhengming's vision of the garden. The little pond has widened into a great lake stretching away to the foothills of some solid yet ethereal-looking mountains on the horizon. The artist has thus made his point right at the beginning: the wall is there, and his scenes will all be inside it. But for the rest of the album the viewer must suspend disbelief.

Flying Rainbow Bridge, Zhuo Zheng Yuan, Suzhou, as it is today. The garden has been totally rebuilt since Wen Zhengming's time, although an ancient wisteria he is said to have planted still existed in the early years of this century.

Elaborate Gardens

ABOVE, LEFT *Xiao Fei Hong (or Little Flying Rainbow Bridge), album leaf of the Zhuo Zheng Yuan, Suzhou, by Wen Zhengming, 1533. A second album of paintings of this garden. 'Bright is the moon,' says the poem to this picture, 'and oh, how far-reaching is the sky!'*

ABOVE, RIGHT *Zhuo Zheng Yuan, Suzhou. Although the city streets have been shut out, the distant pagoda – 'borrowed' into the garden visually – extends its boundaries beyond the enclosing walls.*

The Zhuo Zheng Yuan still exists. To the Chinese way of thinking it is one of the most beautiful of all the gardens that remain in China, as much because of its association with Wen Zhengming as for its own qualities. However, a visitor who knows the album leaves will be surprised by its present appearance. For in Wen Zhengming's pictures the garden is a rustic retreat, in essence still a scholar's hut, though on an enlarged scale; whereas today it is one of the best surviving examples of the elaborate garden tradition, which stems less from Wang Wei and Bai Juyi than from Shi Chong and the great summer palaces of the Emperors. Instead of simple thatched pavilions and rustic lodges, the visitor is led by degrees into a sophisticated maze of fanciful gazebos, twisting galleries, courtyards and formal halls, all surrounding a central pond. The garden is famous for its fine workmanship, for its complex succession of views and its elaborate elegance.

The visual conventions for this elaborate tradition were partly developed by a painter who was not an amateur *wenren* artist but a highly skilled professional. His name was Qiu Ying, (*c.* 1494– *c.* 1552) and for fanciful garden design he may be the most important painter of all. His work, like Wang Wei's scroll, expressed a garden ideal, but he drew on features which in his day were already well-established conventions of garden design. However, he composed his gardens so beautifully and peopled them with such exquisite ladies and elegant gentlemen that they must have been irresistible. To give his garden paintings background and weight, he often set them back in the classic periods of Chinese history. Any rich man seeing such a garden could hardly have failed to want one like it. Qiu Ying's paintings were in several styles, but it was his historical fantasies, in the blue-and-green style, which were most admired. They sparked off endless imitations and not a few forgeries, on account of their consummate grace and elegance. We have already looked in detail at one of his scrolls, showing the garden of Sima Guang, which was painted with a classic, spare refinement appropriate to this garden of an official in voluntary retirement. Osvald Sirén has already analysed, in detail, a garden scroll in the style of Qiu Ying;[14] illustrated below is a painting in a similarly precise manner which captures the feeling of the elaborate and elegant garden tradition. In it we can clearly see the variety and plenitude of a grand city garden. Walls dissolving into mist, formal hall and informal study, Taihu stones and miniature mountains, tortured pine and blossoming plum, underground grotto and two-storey *lou* giving a view of the hills – nearly every garden element is rendered here in a crisp elegance which would have flattered the owner. No wonder such representations have appeared on screens, porcelain and lacquerwork ever since Qiu Ying and his professional followers formulated this classic mode.

Convention and Excess

In the eighteenth century, when the amateur ideal had become too self-conscious for its own good, the whole problem of gardens as expressions of character was given an interesting new slant. In a book on gentlemanly pursuits by Li Yu (whose pen-name was Liweng, 'the Old Man in a Coolie Hat'),[15] the grace or vulgarity of a rock garden is still seen conventionally as the direct result of the owner's own spirit. However, for Li Yu, the skilled mountain builder was likely to be not the scholar or painter or poet, but the much despised artisan. Rock artisans, he writes, have throughout history produced by far the most beautiful compositions, in spite of their narrow specialization in particular skills. The reason for this is that they are close to nature and their manner is instinctive, while painters and poets have had too much training. In rock building, as the *Yuan Ye* had already pointed out, he who tries too self-consciously hard will fail: artificiality will dominate and all will be lost. Thus, it is best for the owner to act as overseer, setting high

The Zhan Garden, Yangzhou, anon., nineteenth century. An elegant garden with formal pavilions and straight-angled galleries set among extravagantly informal rocks. The white mist lying along the horizon half dissolves the garden's boundary wall so that it seems as if the distant mountains are part of the layout. It is interesting to compare this with the painting on pages 96–7, where a simple garden is painted with similar delicacy of detail.

standards and accepting only the most refined and sensitive work from the workmen whom he has employed to transfer the *qi* of nature directly to the garden. In a sense, the artisan becomes the brush through which the owner's spirit and that of nature find perfect expression in the garden's construction.

Such an explanation, however, does not take account of the inordinate amount of time needed to create a garden. Gardens can never be really finished; they are a constant reflection of the owner's care and interest, as well as his character. This might seem one of the reasons why the great Yuan painter Huang Gongwang (1269–1354) had been so constantly held up as the model for garden makers. For his method of painting was a slow, additive one whereby he would lay out a composition when the inspiration was right, and then spend as much as three years filling it in as the mood struck him. Clearly this manner of working was much more applicable to gardens than the swiftly completed brushstrokes of bamboo painting, but it seems more likely that it was the look of Huang Gongwang's mountains that was recommended, rather than his method of creating them. In one of his few paintings to have survived, *Dwelling in the Fuchun Mountains,* the hills are portrayed in an impressionistic manner, by a series of dark ink-strokes dotted evenly around a white background. They give the feeling of an overcast, wintry day. The style is self-consciously sketchy, lacking the kind of elevated sentiment which would have appealed to the 'vulgar'. In fact Huang Gongwang seems to lack even the obsessive purity of Ni Zan's sparse vision. He was celebrated for 'plainness', and a contemporary reached the zenith of success by achieving 'flavour with blandness'. These qualities are a long way from the 'divine' mountainscape of the northern Song, the dramatic 'lyric' of the southern Song, or the elegant refinements of Qiu Ying. The mountains are unimposing hills, the dwellings nestle in the valley, everything is modest, informal, straightforward. What, therefore, was the *Yuan Ye* recommending

to its readers when it described a rockery as 'built up of rugged stone blocks jagged and split as in (Huang Gongwang's) paintings'? It is hard to guess for such a description does not seem to fit the master's vision of mountains in the *Fuchun Mountains* scroll, where he seems to be aiming at a generalized portrayal of 'mountainness' rather than at any more dramatic concentration of forces.

Gardeners too, of course, tried to capture this same feeling of the essential – we are back again at the *qi yun* – but evidence from painting suggests that they were achieving it more and more through sheer profusion of detail, something which Huang Gongwang specifically warned against. As time went on rockery mountains threatened increasingly to monopolize all the garden space, and they are much less like the cool, generalized hills of Huang Gongwang than the writhing, expressionistic rock forms of his slightly later contemporary, Wang Meng (1308–1385). A late forgery of Wang Meng's style shows great contorted Taihu rocks dominating the entire character of the garden. Arranged in the recommended way near water, they are built up to twice the height of the small halls, and their grotesque shapes loom over the roofs like mushroom clouds. They remind one of Guo Xi's gigantic mountainscapes, where bizarre faces bubble out of cliffs and peaks, but they lack Guo Xi's dynamic sense of life and movement and though decorative, lie flatly on the page. It is possible that the lifelessness of the painted rocks reflected those in the garden itself.

'Landscape,' a forgery in the style of Wang Meng, (perhaps by Wen Boren, sixteenth century). Taihu rocks dominate this garden, which aims to re-create the magical mountains of the Immortals rather than a naturalistic landscape. The extravagance of the rock-work, however, is balanced by the comparatively modest designs of the buildings. The garden is thus a transitional one, halfway between the elaborate-Imperial-Immortals'-paradise tradition, and that of the modest 'scholar's retreat'. Note the crane, a symbol of longevity.

In the sixteenth century, which is when this forgery may have been painted, critics were upset by the decline they felt had taken over in painting. One of the worst mistakes, said the painter Mo Shilong, was the way in which 'people build up small bits to make a large mountain'.[16] The *Yuan Ye* and the writer Li Yu were both to point out the same fault in the building of garden rockeries. In late gardens it is easy to see what provoked these complaints; close up, such rock piles can be quite fascinating, but from a distance they are bewildering heaps of stones, stuck together, as the modern connoisseur C. C. Wang expressed it, 'like lumps of peanut candy'.[17]

This decline in overall composition may well have been accelerated by the seventeenth century manuals on painting technique, such as the *Mustard Seed Garden* (1679), which gave innumerable examples of all the individual elements that could be combined in a painting. This kind of instruction encouraged students to concentrate on the details in exactly the way the masters deplored – at the expense of the 'vital spirit' of the whole.

Nevertheless, from the end of the Ming period there was a fresh attempt at an aesthetic revival which reached its full development in the early Qing. In the works of the monk and great 'individualist' painter, Dao Ji, one once again feels the artist's own powerful life-force expressed in multitudinous mountainscapes that reveal not the characteristics of particular rocks but the whole power of creation. For Dao Ji, who called himself the 'Monk of the Bitter Cucumber', the emotional force of the landscape was expressed 'by making some parts in them wide open and other parts hidden or screened'; for gardens, whose designers were striving for more and more effects in small spaces, this would become an important principle. Nevertheless, in spite of this emphasis, Dao Ji insisted on a return to detailed observation of nature, trusting in his own inborn genius instead of following in the shoes of the antique masters. For him what mattered beyond this, however, was that the whole soul of the painter must be in his work: 'No matter what the

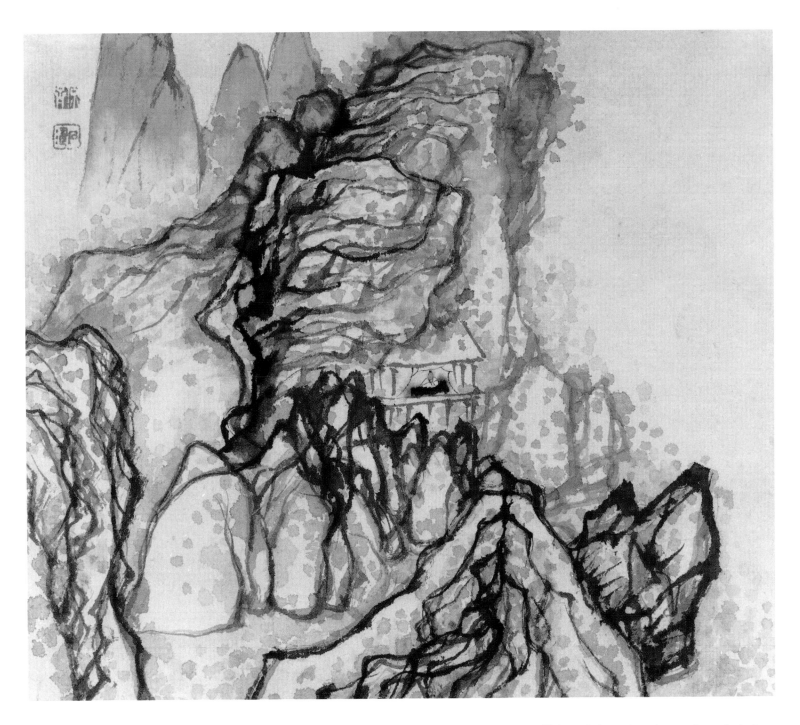

'A man in a house beneath a cliff', by Dao Ji (1641–1717). James Cahill says of this: '[Dao Ji] is not so much depicting rocks as presenting to our senses the forces that mould and destroy rocks. Experiencing empathetically the movements of his hand as he wielded the brush, we take part in an awesome act of creation'. Although the cliff seems about to eat up the hermit in its monstrous jaws, he remains unperturbed, at one with the forces of creation. As well as a remarkable painter, Dao Ji was a well-known rock-builder, and spent several years creating 'false mountains'.

paintings look like: if they only express the essence of life and the painter's creative mind, they are examples of the all-inclusive creative art'.[18]

But in the hands of less powerful and brilliant men such views, rather watered down, could be unfortunate. Without Dao Ji's dynamic force and discipline, it was easy to explain away extravagant but lifeless work as 'personal expression'. In such cases excess, as in the Tiger Balm gardens in Hong Kong and Singapore, often takes over from inspiration. Instead of going back to nature to understand the forces which generate mountains, certain builders of rockeries have tried to invigorate old traditions by exaggerating the forms to extraordinary lengths. The true way, however, has always been to capture the *qi yun*. For in a successful garden, forms themselves are only starting points for the imagination to roam freely about in the haunts of the Immortals, the rustic huts of great sages and scholars, and the minds of those masters of landscape who had immortalized themselves in painted mountains.

ARCHITECTURE IN GARDENS

In the second quarter of the seventeenth century a sympathetic gentleman scholar named Qi Biaojia (1603–45) suddenly found himself obsessed by the making of a garden. 'In the beginning,' he wrote,

> I wanted to build only four or five rooms, and some friends told me where I should build a pavilion and where I should build a summer house. I did not think seriously of these suggestions, but after a while these ideas would not let me alone, and it seemed indeed I should have a pavilion here and a summer house there. Before I had finished the first stage, new ideas forced themselves upon me, and they chased after me in all out-of-the-way places, and sometimes they came to me in my dreams, and a new vista opened before my imagination. Hence my interest grew more and more intense every day and I would go to the garden early in the morning and come back late at night. . . . Early in the morning, while resting on my pillow, I saw the first rays of the morn and got up and asked my servant to go with me on a boat, and although it was only a mile off, I was impatient to get to the place. This continued through winter and summer, rain or shine, and neither the biting cold nor the scorching sun could restrain me from it, for there was not a single day when I was not out on the spot. Then I felt under my pillow, and knew my money was gone, and I felt annoyed over it. But when I arrived at the spot, I wanted always more and more stones and material. There are two halls, three pavilions, four corridors, two towers and three embankments. . . . In general, where there is too much space I put in a thing; where it is too crowded I take away a thing; where things cluster together I spread them out; where the arrangement is too diffuse I tighten it a bit; where it is difficult to walk upon I level it; and where it is level I introduce a little unevenness. It is like a good doctor curing a patient, using both nourishing and excitative medicines, or like a good general in the field, using both normal and surprise tactics. Again it is like a master painter at his work, not allowing a single dead stroke, or like a great writer writing essays, not permitting a single unharmonious sentence. . . .'[1]

Two aspects of this account are especially striking. First, the gentleman-scholar's garden is not attached to his house; and secondly, his interest is all in architecture and layout, and nowhere in his description does he mention flowers, trees or shrubs.

In England and America we plant a garden, but a Chinese garden is built. A Western visitor may be astonished by the contorted piles of rock in Chinese gardens, but he will be no less amazed by the extraordinary number of buildings. Written descriptions of the various features of a garden give the impression that each pavilion is separated from the next by wide expanses of open scenery, or by acres of mountains and trees, and it is true that this was the desired effect. In city gardens, however, where space was at a premium, an extraordinary number of buildings and rockeries were often crammed into gardens of no more than an acre. Even the large gardens of immensely rich families favoured tightly packed spaces over wide and expansive views.

A drawing of the Jia family gardens, from the eighteenth-century novel *Dream of the Red Chamber* (*The Story of the Stone*), shows a startling number and variety of buildings. No two are

alike and they are all jumbled in among rocks and water like a setting for some fantastic obstacle race. Moreover the garden is equally full of people – as indeed it is in the novel itself. At one point in the *Dream of the Red Chamber* the garden is inhabited not only by the young master of the household and all his sisters and cousins, but by all their attendant maids and serving girls, a troop of child actresses and even a community of youthful Buddhist nuns. Such a very large number of inhabitants was perhaps unusual even for an extremely large garden, nevertheless garden pavilions were often lived in – not only in summer, but throughout the year. Sometimes a neglected garden might harbour only a caretaker and his wife, but in many cases a resident scholar, an old family tutor or friend down on his luck would live in one of the studios and make use of the library – an essential feature of any garden worth the name. Rich men also often housed their lesser wives and concubines in garden pavilions, as in the erotic novel *Golden Lotus,* whose heroine, momentarily neglected by her lord, has a romance with one of the young gardeners.

In fact the individual rooms that go to make up the Chinese house are essentially no different from those in the garden. Despite their fanciful shapes, Chinese garden buildings, with their posts and beams supporting tiled roofs on cantilevered brackets, share the same basic structure with the most prosaic buildings of the house. And since there are no rooms in the house specifically shaped or decorated for dining rather than bathing – an adequate kitchen may be set up in any room by bringing in a charcoal brazier, or a bedroom by adding a sleeping dais – a garden pavilion may easily be converted into a dwelling which has just the same degree of comfort as a set of rooms in the house.

ABOVE   *Da Guan Yuan (Prospect Garden), an imaginary reconstruction of the garden in the novel* Dream of the Red Chamber (The Story of the Stone), *from* Zeng ping bu tu Shitou ji, *Shanghai 1930. Extravagant rocks loom over the buildings, which are all slightly different. A stream winds between them, crossed by an equally varied collection of bridges.*

RIGHT, ABOVE   *Dragon-headed wall, Yu Yuan, Shanghai. One of five such heads in the garden. Though the workmanship is very fine, and the heads in their own way splendid, garden connoisseurs disapprove of such literal interpretations since their lack of ambiguity limits the imagination.*

RIGHT   *Three doors in a recently restored garden, Nanking. The line-up of doorways leads the eye through to Mochou lake and the garden beyond. The vase shape at the end is a visual pun: the Chinese for 'vase',* ping, *sounds like the word for 'peace'.*

Nevertheless there are crucial differences. Above all, house buildings are always sited on an orderly rectangular plan, and in large households are arranged as regular progressions of courtyards. A garden, on the other hand, is instantly recognizable because everything in it is irregular and confusing; it is a place where the ordinary is transformed into something new and delightful.

Architecture as Metaphor

In a Chinese garden architecture is more playful than useful and, above all, more metaphorical. Gardens allowed the normal city architecture to be liberated from Confucian rectitude, much as in Western garden architectural follies, historical pastiches and even Chinese-style pagodas represented a loosening of urban decorum. Naturally it was understood that freedom could be misused, and many critics warned garden makers against the dangers of 'vulgarity' – too many whimsical dragon heads. Happily, the warning was interpreted quite freely, for an essential aspect of the Chinese garden is its playful transformation of the animal and vegetable kingdom into architectural forms.

Merely to list some common architectural elements in a garden, with the Chinese phrases that have been associated with them, is to uncover this metaphorical dimension. Holes through a wall can be circular 'moon gates', while sometimes they are in the shape of flowers, shells, gourds or vases. Balustrades can take on the pattern of 'cracked ice', pathways can become 'geese' and 'meander like playing cats', pavilions over the water are 'boats', and five pavilions set together become 'the claws of the Imperial five-toed dragon'. Rocks, of course, are 'goblins and savage beasts', unless they are 'bullet-holes'; a willow tree 'sways like the slender waist of a dancing girl'; and the heart of the garden, the water, is where 'the moon washes its soul'. This delight in metaphor is noticeable even in the plans of buildings: some are in the shape of plum blossoms, or fans – a popular design since they combined the idea of cool breezes with a useful form for linking two galleries round a corner. Sometimes two pavilions are joined together at the corners to form a butterfly; reality is everywhere transformed into a poetic conceit.

Pavilions

In fact the simple presence of a pavilion – one of those free-standing, open-sided, fancifully shaped viewing pavilions which the Chinese call a *ting* effects one of the most fundamental transformations in the garden. 'Once a place has a *ting*,' the saying goes, 'we can call it a

ABOVE *Leaf door, Nei Yuan, Shanghai. An uninspiring corner enlivened by the juxtaposition of a fanciful doorway with a simple window. The door shape is open, the window filled in with a fish-scale pattern.*

LEFT *Marble boat, Shi Zi Lin, Suzhou. A late and not altogether successful addition to the garden, this concrete tea pavilion in the upper right-hand corner fills an area of the lake perhaps better left open. (Drawing in Shi Zi Lin, Suzhou)*

BELOW *Flute player in a* ting *on the edge of a cliff, by Luo Ping (1733–99). Such little, open pavilions placed in nature domesticated the landscape. In gardens they suggested that the carefully placed rocks around them were the work of nature.*

Ladies in a garden pavilion playing chess, by Leng Mei (eighteenth century). The door is petal-shaped, emphasizing feminine grace. In the garden roses have been trained up over a trellis, a popular seasonal screen. For other examples of doors see pages 119, 120 and 139.

garden'. A piece of architecture domesticates a wilderness or, as the Chinese say, 'borrows' the landscape, creating both the frame and the focus it lacked before. The function of habitation plays no part in this, because a *ting* can provide only shade or shelter from light rain. Lacking walls, it is merely a resting place, offering a view to which the cultivated visitor will respond by supplying infinite layers of association. Set in a garden or in nature (a *ting* nestling in the hills is a major theme in painting), it symbolizes man's tiny but essential place in the natural order: just as without man nature lacks a focus, so too a garden landscape can scarcely exist without a *ting*.

However, a city garden can hardly exist without many other types of building too. Closely related to the *ting* is the *xie*, an open pavilion attached to an entrance, like a porch, or jutting out from an open gallery, a little room to rest in on a walk, usually beside water. There is also the quiet, enclosed *zhai*, a study or little studio, and the whimsical *fang*, described by the Chinese as 'a boat with no mooring'.

FAR LEFT *Xie*, lou *and* tang, *south-east corner of lake, Liu Yuan, Suzhou. The small porch-like pavilions on the left could be described as xie; the two-storey pavilion in the centre as a lou, and the elegant hall, just visible on the right, as a tang. The steps leading up to the top floor of the* lou *are hidden in the rocks behind it since staircases were considered unsightly. Garden designers find all kinds of ingenious ways to hide their staircases, and if all else fails, leave them out altogether.*

LEFT, ABOVE *Lake pavilion, Xie Yuan, Suzhou. Double roofs doubled again in reflection. This pavilion contains another pavilion inside it.*

LEFT, BELOW *Interior of Hall (or* tang*) of Distant Fragrance, Zhuo Zheng Yuan, Suzhou. This elegant hall is kept cool and dark by the covered gallery that runs around it outside the windows. The rectangular arrangement of furniture is characteristic in such formal* tang.

ABOVE, LEFT *Double-tiered roof, main hall, Yu Yuan, Shanghai. Elongated eaves swoop out in a complex symphony of curves. An elephant tops one cluster, water monsters curl their tails above the main roof beam.*

ABOVE, RIGHT *Greenhouse, hotel garden, Wuxi. The only example known to the author of a Chinese roof in glass.*

Apart from these little pavilion types, gardens also included much larger constructions like the multi-storeyed *lou* or *ge*, and grand reception halls or *tang* which occupied more central and public parts of the enclosures. Between one type and another, however, there was always a considerable overlapping of functions. And the structural aspects of all these buildings seldom varied from the basic beam-framing systems, and from a limited number of roof forms which garden builders excelled in elaborating.

Roofs are not treated separately in the *Yuan Ye* or any other Chinese book on gardens, yet their visual prominence, so unusual to foreigners, seems to demand special attention. They are seen looming large over walls and through trees, and often their sweeping forms totally dominate the pavilions which support them. In short, the curving up-turned roof is a primary expressive element; experiments with flat-roofed summer houses in modern gardens have been disastrous. Some roofs are so disproportionately large that the sheltered enclosure below seems a mere afterthought, an excuse to display the gentle – yet at the same time awesome and powerful – rush to the sky. It is often said that the up-turned Chinese roof appears to float above its recessed base; certainly it is lightened by the smiling curves at each corner. However, there is also a counter-force to this effect, for the roof also seems to brood over the space it covers. Thus it is another instance of ambiguity; partly a light handkerchief hovering miraculously over shadowed columns and brackets, partly a unifying mass which caps and dominates all other elements.

The Site: Improving on Nature

How were all these pavilions to be arranged in a garden? The author of the *Yuan Ye* states – quite boldly in the circumstances – that there are 'no definite rules for the planning of gardens'. He thus confronts himself with a paradox: how to write rules without writing rules, or how to communicate the principles of creating a garden without making them lifeless prescriptions. He does this, as painters did, by referring to the *qi yun*, the life-spirit, which should lie behind all creative work, and also by hiding his prescriptions in a series of poetic images, *aperçus,* and even contradictions. These last are especially useful because they allow freedom of choice. Thus at one point the manual recommends that 'one should preferably not plan a garden in the city' (because of the noise), but at another suggests that the city is preferable to a country site because it is more comfortable and accessible. In other places the *Yuan Ye* avoids rules altogether, as we have seen in the painting chapter, by the use of that typically Chinese technique – evocative description. By describing the emotional effect of a beautiful scene, the author creates in his reader the desire to make a garden that will trigger the same responses – leaving the details for him to think out himself.

Again and again the garden builder is urged to avoid the obvious, and to seek to obtain the unusual and unexpected. Secret places are especially desirable, for they are some of nature's most delightful effects. A wall is to be 'hidden by creepers', buildings should be 'partly concealed by trees', and there should be, here and there, a 'sequestered spot'. But from these intimate places it is important that one should be able 'to gaze far away, as over endless waters'. The key words here are 'as over'. All gardens are based on illusion, and what must be captured is the effect of infinitude – even within the confines of the city. This concept has been most fully realized in the garden-cities of Jiangsu and Zhejiang provinces, in that area which the Chinese call Jiangnan, 'South of the Yangtze'. The great gardens of the north tended to be laid out in more formal patterns. Those that can still be visited in Peking have a certain stiffness compared to the swoop and irregularity of the southern gardens, a more sober and grand feeling which seems to suit the greater formality of the imperial capital, as well as the more rigorous climate of the north. Nevertheless, the emphasis on meandering through a maze of complex and continuous surprises is common to all Chinese gardens. In particularly elaborate gardens much of the surprise comes from the disposition of architectural features. However, even those gardens which are crammed full of walls and buildings still aim always to make the manmade elements blend into and not dominate the natural ones. In a sense this principle is the opposite of that which underlies Le Nôtre's style of gardening as practised at Versailles. While the French treat nature as if it were architecture, planting trees in avenues and clipping hedges into walls, the Chinese try to make the many architectural elements of their garden conform to an ideal of natural irregularity.

Some Chinese authors connect symmetry in a garden with over-design and vulgar elaboration. Symmetrical arrangements lead to stiffness and conformity, and so lack that sense of pleasant and unforced accumulation which ideally makes the buildings seem to have grown of their own accord – like flowers – on the site. Of course accumulation itself can be carried too far. Although a garden should ultimately be a labyrinth, it should not obviously look like one. When the Jesuit fathers in Peking devised a Western maze for the Qianlong Emperor, it was regarded as a barbarian novelty, and soon fell into disrepair after the Emperor's death. The maze, of course, was too regular, too false, whereas a true Chinese labyrinth garden should always seem spontaneous and uncontrived.

This requirement corresponds to the Western notion of the picturesque, suggesting Alexander Pope's famous advice 'Consult the genius of the place in all'. In China, as in eighteenth-century England, a good garden maker first carefully assessed the natural contours and water levels of his site, exaggerated them where necessary to intensify their effect, and then designed his garden buildings to fit into this modified landscape. 'Naturalness' thus consisted in making the whole thing look as if it had happened without human aid. Tong Jun and other authors perpetually point out that this quality was rarer after the Ming dynasty, when the elements of garden and architecture became more formalized and repetitive. Rocks and flowers, they complain, were laid out like soldiers; and brackets under eaves looked like a file of resplendent cavalry on parade. Such regularizations, 'which have a Latin feeling', fall very short of the ideal of natural spontaneity which had been attained in the Song dynasty.[2]

Plan of the Yu Hua Yuan (back garden, Imperial City, Peking). This garden is a strange, stiff, symmetrical place full of piles of contorted rocks and paved all over with pebble designs. It is, however, still magical, on a hot spring morning, to sit on the marble terrace (1) among the sighing, ancient juniper trees, and look down on people quietly wandering along the shaded paths below.

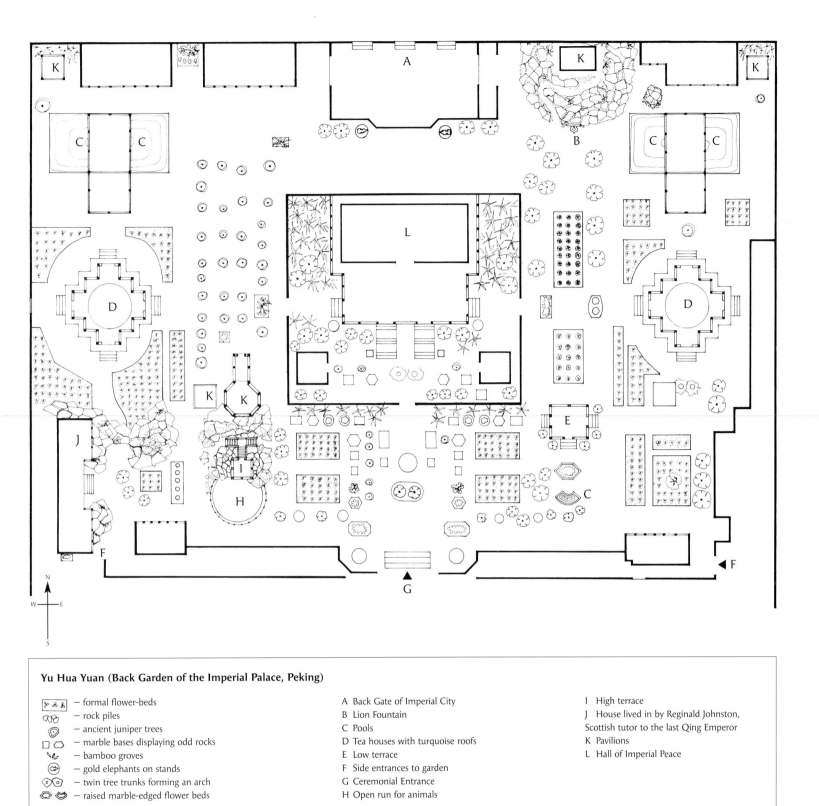

Yu Hua Yuan (Back Garden of the Imperial Palace, Peking)

- 🌸 – formal flower-beds
- 🪨 – rock piles
- ◎ – ancient juniper trees
- □ ◯ – marble bases displaying odd rocks
- 🌿 – bamboo groves
- ↻ – gold elephants on stands
- ◎◎ – twin tree trunks forming an arch
- ◇◇ – raised marble-edged flower beds

A Back Gate of Imperial City
B Lion Fountain
C Pools
D Tea houses with turquoise roofs
E Low terrace
F Side entrances to garden
G Ceremonial Entrance
H Open run for animals

I High terrace
J House lived in by Reginald Johnston, Scottish tutor to the last Qing Emperor
K Pavilions
L Hall of Imperial Peace

The Yu Yuan

One of the problems with gardens is that they are vulnerable not only to the ravages of time, but also to the new ideas and restorations of owners not always sympathetic to the original maker's intention. In China today no gardens still exist in their original form, and what the visitor sees as a Song or Ming garden inevitably includes a great deal of Qing and even more recent building. The Yu Yuan in Shanghai, for instance, is a late Ming garden which, because of later

137

LEFT *Complex spaces: view along northern boundary wall, Yu Yuan, Shanghai. The beautiful roof below belongs to the Ming dynasty Cui Xiu Tang, or Hall of Fine Things Gathered Together. This part of the garden makes good use of the principles 'small leads to large, high to low'.*

RIGHT *Internal wall, open gallery, xie and ting, all edging the main lake, Zhuo Zheng Yuan, Suzhou. See page 183 for a plan of this garden.*

RIGHT, BELOW *Corner of the central lake, Wang Shi Yuan, Suzhou. The different buildings are connected in an extraordinary variety of ways, yet the whole effect is harmonious and balanced.*

additions, the *Yuan Ye* would very likely condemn as exceedingly vulgar. Nevertheless it is a fascinating architectural labyrinth, and contains almost all the delightful and surprising elements of a city garden. In plan it is a sequence of five major areas divided by long, snaking walls, which undulate through the garden in regular curves to end in dragon heads. Within the five sections there are altogether fifty or sixty small, discreet space cells; tiny places which give the 'unexpected views' demanded by the *Yuan Ye*.

Unlike Western architects who might try to expand tight spaces by joining them together, the Chinese designers increased the *feeling* of space by dividing it up with screens and walls. By layering the available space with gateways, by allowing glimpses of swooping rooftops and patches of light, they managed to suggest that space extended infinitely and magically beyond

LEFT *Half of the interior in the Hall of the Veteran Hermit Scholar, Liu Yuan, Suzhou, a hall divided into two equal parts. The finely carved windows bring in light from a minute open corridor behind them, but distract the eye from discovering how cramped the space is there. To the north, full length windows open on to a large enclosure with pavilions and a pool.*

its visible confines; for the aim was always to represent nature's infinite change and mystery as well as to provide seclusion.

In most gardens the space cells unfold towards and around a central 'mountain-and-water' area, protecting it and giving it a sense of introversion, as if one had to peel back layer after layer of building to reach the core. At the Yu Yuan these cells were added in a succession of building campaigns across time. Rocks and water created the garden first, with a pavilion and bamboos. Pan En, who originally laid it out in 1559, and whose description survives, does not mention an overall plan, nor is there any reference to walls. The garden is simply begun by assembling the essential elements according to spontaneous feeling:

> For twenty years I continued to build the garden. I sat a sit – I thought a thought – I rested a rest – it was still not very good. In [the year 1577] ... I gave my entire heart to the affair. I thought only of the garden. I increased the size of the ground adding fifteen plots of land. I made seventeen pools. Furthermore I bought many fields and devoted the entire revenue from these to beautifying the garden. [3]

In other words this gentleman-official was adding to his garden piecemeal in that almost obsessional way recorded also by Qi Biaojia whose experiences are quoted at the beginning of this chapter. Both men allowed gardening to become a crucial part of their lives, and their gardens were therefore a kind of autobiographical record, reflecting their fortunes and moods. Pan En continues his account:

> To the east I placed several two-storeyed buildings in order to cut off the noise and bustle of the city and in the centre of the three divisions I made a doorway over which a board was placed upon which the characters 'Yu Yuan' (Garden of Ease) were written. Entering the gate, taking a few steps to the west there was again a gate, on this was written 'Jian Jia' (Beauty Penetrates Gradually); twenty paces more to the west and one turned to the north, at this place there was a small *paifang*

RIGHT *South-west corner, first courtyard, Yu Yuan, Shanghai. A covered gallery, with intricate lattice work windows, surrounds the first hall. This is a corner of the hall illustrated on page 135.*

BELOW *Sea monster swallowing roof rib, Nei Yuan, Shanghai. Animals and heroes on horseback populate the roofscapes of both the Nei and the Yu Yuan in Shanghai.*

(an arch), on which were inscribed the four characters 'Ren jing hu tian' (Man's Place is in the Immortal's Heaven). . . . Following the wall to the east one reached a hall called the Jade Glory and in front of this stood a marvellous rock, a most curious example of Nature's handiwork. It had been known and remarked since the time of the Emperor Huizong of Song, and should by rights have been sent to him, as it was considered the finest in the Empire. The name of the rock was Yu Ling Long (Exquisite Jade), and the hall took its title from the rock.

Today the garden still contains a large Taihu stone with a name similar to that of Pan En's (although it is hard to tell if it actually is the same stone), and there are many fine rockeries round pools, as well as the huge artficial mountain that dwarfs the buildings of the first and second areas. It is noteworthy that Pan En names one of the rockeries the 'Immortal's Heaven',

but concludes his whole description with a comparison to that most famous of all early poet's retreats – the riverside estate of Wang Wei.

According to the records the people of Shanghai bought this garden from Pan En's descendants in 1761 and repaired it with public funds, calling it the Western Garden. At that time it consisted of just under thirteen acres (77 *mu*) and remained in good condition under the subsequent ownership of the Bean Merchants' Guild. Unfortunately, during the Taiping rebellion in 1861, Western soldiers were housed in the garden, and to provide extra ground for their accommodation the rock hills were dumped into many of the pools. Since then it has become difficult to determine exactly which of the features of the garden correspond to passages in Pan En's description. The garden was restored in 1956, and now includes the Inner Garden or Nei Yuan – an extraordinary over-complicated tangle of interwoven elements – as part of the same complex.

The first space cell is reached by entry through the main gate (A) and

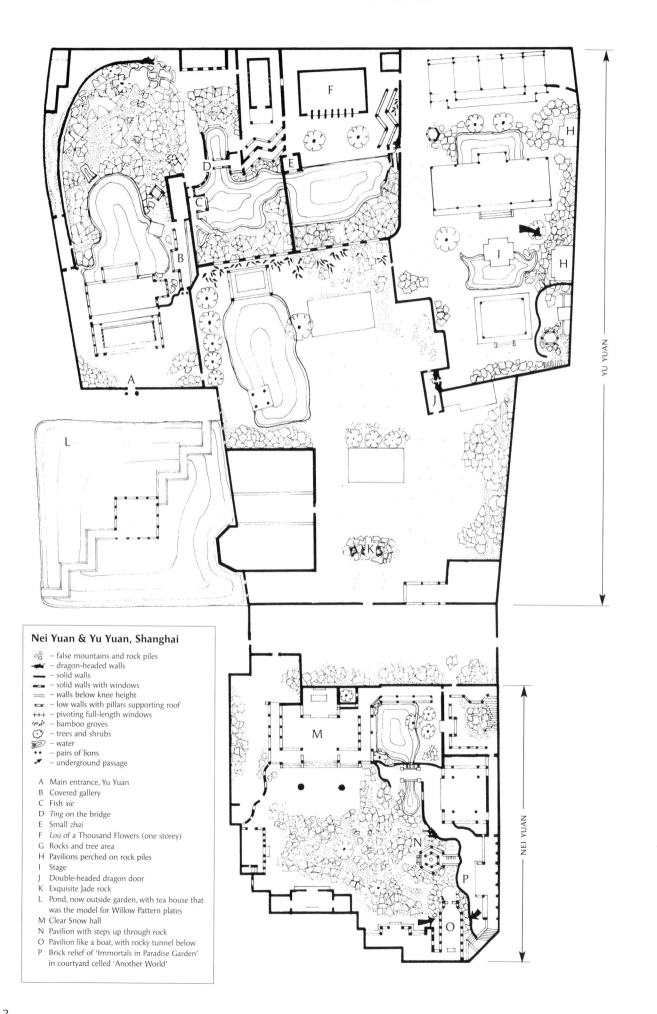

Nei Yuan & Yu Yuan, Shanghai

- – false mountains and rock piles
- – dragon-headed walls
- – solid walls
- – solid walls with windows
- – walls below knee height
- – low walls with pillars supporting roof
+++ – pivoting full-length windows
- – bamboo groves
- – trees and shrubs
- – water
- – pairs of lions
- – underground passage

A Main entrance, Yu Yuan
B Covered gallery
C Fish *xie*
D *Ting* on the bridge
E Small *zhai*
F *Lou* of a Thousand Flowers (one storey)
G Rocks and tree area
H Pavilions perched on rock piles
I Stage
J Double-headed dragon door
K Exquisite Jade rock
L Pond, now outside garden, with tea house that
   was the model for Willow Pattern plates
M Clear Snow hall
N Pavilion with steps up through rock
O Pavilion like a boat, with rocky tunnel below
P Brick relief of 'Immortals in Paradise Garden'
   in courtyard celled 'Another World'

either past or through two large halls which are placed close together and face south. These dignified and impressive halls (public reception areas and viewing pavilion) are surrounded by raised galleries of red-lacquer pillars marching in step, and fanciful windows in which are carved stylized cranes and plum blossoms. Behind them is the pool and, pressing in on it, an extraordinarily tight and high rock mountain. The feeling in this space is at once peaceful and dynamic: the powerful rocks and complicated buildings crowd up against each other and loom over the small pool, dwarfing it into a narrow, shady canyon – yet one can still lean over the side of a balustraded seat and admire the calm water (B). From this first cell there are glimpses of further spaces and more vegetation to come; most notably, some bamboo can be seen over the tiled rooftops, and several very high metasequoias and bright green willows swinging gently in the breeze. The rest of the garden lies to the east. To get there one traverses a very tight set of minor space cells. Here there is a sense of confusion and expectancy. Further on, to the right is a view through the Fish or Happiness *xie* (water pavilion) (C), and straight ahead through a door there is a covered bridge (D) to another pavilion. This later becomes a gallery that divides into two parts which zig-zag back and forth in unison. The wall separating these two galleries is broken every few feet by small, shaped openings, so that one can see through the division to the path not taken.

This double gallery leads past a little study pavilion overlooking the water (E), and from there to an open space in front of a large, two-storey hall, the *Lou* of a Thousand Flowers (F). The space enlarges into a terrace, creating an effect which is relaxed and spacious in contrast to what has gone before, but at the same time more intimate. Two old trees are planted here, a four-hundred-year-old ginkgo and a hundred-year-old magnolia. Together they shade the entire courtyard, which in summer is filled with the scent of magnolia blossoms. Standing beneath them, facing away from the hall, one looks across a little rocky-banked stream to more rocks, where in springtime camellia trees are placed in pots (G), making bright splashes of scarlet and pink among the stones. Any further view is blocked by a white wall roofed in grey tile which acts restfully upon the mind, like mist over a lake, or the pale silk background of a landscape painting. Next, one proceeds through a double-walled gallery and through a dragon wall to the third main section of the garden, where three large halls and a stage are arranged north to south with ponds set between them. To the east of this space, which borders the exterior road, there are stone mountains built to a height of three storeys in such a way that pavilions (H) seem to grow directly out of them at the top. The stage (I) is placed over water, which gushes up from a carved stone fountain placed between its pillars. Above all the pavilions the upturned tails of dolphins play on the roof ridges. The western wall of this enclosure writhes and falls and zig-zags in echelons until it ends in two fearsome dragon heads confronting each other (J).

The fourth section, owing to many demolitions, has the most open space within it. For a foreigner – even one sympathetic to the fullness of a Chinese garden – this is a pleasantly relaxed part of the Yu Yuan, for it unintentionally suggests the wilder, more open aspects of nature. Here are the rustling camphors and grove of metasequoias which have revealed their tops over the

garden wall previously, and here too is the pride of the garden, the Exquisite Jade Stone (K), with the Jade Grace Pavilion placed nearby for viewing it. This area also contains a place for growing plants and, behind a wall, two fine halls and a courtyard bordering on Pan En's old pond (L), now outside the garden enclosure.

The Nei Yuan lies beyond this, providing quite a different feeling from all that has come previously. This is the most sheltered and intense series of spaces in the garden – even the decoration is compacted. The small enclosed areas are dark and dank at certain points, thus forcing the visitor into a closer view of details, but in the centre there is one relatively large space which relieves all this pressure. Here is a grand but peaceful hall (M), which opens one way to a pool shaded by two oleanders, the other way to a rockery framed by two snarling stone lions. What the garden lacks in width it makes up in height: a pavilion is piled on tops of rocks (N) and there is even a boat-shaped tea house (O) poised high up over a rocky tunnel, like Noah's Ark after the floods receded. This part of the garden is full of secrets. Even the detailed brick reliefs for which the Nei Yuan is celebrated (P) are often hidden away in tiny courtyards.

Tong Jun, writing in the 1930s, speaks of spaces which are 'tight but not crowded', where one can see a 'void in solids, solids in a void', and where, in general, the 'small leads to big, low to high'.[4]

LEFT, ABOVE   Entrance, Nei Yuan, Shanghai. This small garden, built in 1709, is characterized by many high buildings and small, cramped, winding spaces. Once the visitor grows accustomed to the scale, its complexity becomes fascinating.

LEFT   Hill-top pavilion, seen from outside the Yu Yuan, Shanghai. Inside the garden, a tunnel leads through the rocks below this pavilion.

ABOVE   Brick relief, Nei Yuan, Shanghai. The relief is sunk into the wall of an inner courtyard named 'Another World', and shows Immortals in a garden. The craftsmanship of this and other brick sculpture in the Yu and Nei Yuan is remarkably fine.

These are the sort of effects the designer has aimed at in the Nei Yuan, and indeed they constitute one of the most effective dramatic techniques used in all Chinese gardens. Like a steady rhythmical beat in music they can increase the tempo of discovery and excitement – but the device also operates on innumerable other levels at the same time, some perceived only subconsciously.

In particular the paired opposites, like white wall and dark coping, open pool and rocky tunnel, are an essential expression of the duality of yin and yang, those much-quoted negative and positive principles which, at least from the fourth century BC, have been so fundamental to the Chinese world view. For occidentals, yin and yang are perhaps easiest thought of as the two opposite ends of a pendulum swing – for yin forever gives way to yang, and yang to yin, in an eternal oscillation between the two poles. Thus the yin moon already at its zenith begins to move towards the yang sun, night towards day, winter to summer, cold to heat, soft to hard, dark to light: the possible pairs are not only limitless, but swing at an infinitely possible variety of overlapping rates and tempos. In a garden one could sit all day contemplating these metaphysical movements; some slow, some fast; some obvious, some scarcely felt; some consciously worked out by the designer, some beautifully visible, perhaps by chance – but all of them making manifest, through the senses, the fundamental principles of the universe.

145

Walls, Gateways and Windows

Nowhere is the principle of paired opposites more evident than in the relation of a white wall to whatever is placed in front of it, whether this be tiles or gateways, rocks or plants. After rocks and water the wall plays the most critical role in the garden, for it is not only the most common device for separating different areas, but also provides calm and harmony, serving as a backdrop for the vibrating shadows and silhouettes of bamboo or plum trees. With so many surprising angular shapes competing for our attention, it is relaxing to have this note of stability, this constant unifying theme running all through the garden. Sirén speaks of the white wall as a 'monumental feature' introduced into a medley of otherwise picturesque elements, and he also points out its obvious defensive and heroic overtones. *Cheng*, meaning wall, also came to mean 'city' and the Great Wall is often regarded as *the* symbol of China.

According to the *Yuan Ye*, white walls were basically made of 'earth stamped between boards', and then plastered. 'Connoisseurs who wished to give . . . a glossy surface, used for this purpose white wax, which they rubbed or patted into the wall.' The look of such a wall will change radically through the day. In the morning mist it may disappear entirely leaving only its roofing of dark tiles floating above the ground. Because walls should ideally follow the contours of the site, these tiles may literally give the effect of a 'flying dragon', seeming to wind miraculously through the air with the regular undulations of a watersnake. In other lights, however, the same walls can seem as solid as mountains, and indeed often took like a distant range of hills rising behind the garden rocks and shrubs that are arranged before them. As the sun shifts, different patterns are cast upon the wall, changing again its feeling of solidity and depth. At sunset the mood may become suffused with the melancholy that is characteristic of the more evocative and elliptical style of landscape art. Indeed the most popular interpretation of these walls is that they

Undulating wall and straight boundary wall, Yu Yuan, Shanghai. The inner wall surrounds the garden's huge false mountain, and ends in a ferocious dragon head. A narrow walk slips between the two walls at ground level – some twelve feet below this view.

are to the garden what the unpainted areas of silk are to a landscape painting – not merely a background, but an evocation of infinity. Thus the wall which encloses and divides space also serves to extend it symbolically beyond all bounds.

At times a glossy wall might almost become a mirror. At others the gardener might want an opaque wall, an unobtrusive boundary which would be scarcely visible behind trees and rocks. In such cases he would paint it a soft, blurred grey, which would merge into the stones. A circular hole cut in this kind of wall produces a spectacular effect on moonlit nights: the wall disappears in shadow, while the shape of the circle is cast obliquely on the ground before it – like the moon itself.

Painted and polished walls were not the only way of dividing space. Li Yu, the eccentric seventeenth-century writer, saw a special appeal in walls made of rubble, or of 'plaited bamboo or branches of the jujube bush' – often decorated with climbing roses. The *Yuan Ye* felt that 'such wattled walls are better than trellises; they are more rustic in appearance, and have a fragrance of woods and mountains'. The walls of garden pavilions, of course, could also play a highly decorative role. Although *ting* pavilions were open on all sides, the space between the pillars of other garden buildings was filled in with pivoting windows made of wooden latticework backed by opaque white paper. From inside in the daytime the outlines of the lattice appeared in silhouette, while outside their complex, geometric patterns set up delicate relationships with the shapes of twigs and branches. By night, the effect was reversed. Outside, the silhouettes of lit-up windows made decorative patterns on the darkness, while inside the surface of the latticework itself would be illuminated in contrast to the dark spaces between.

Walls played such a crucial role in China that one is tempted to speak of them as the subject of a kind of mania. Walls ten to twenty feet high surrounded residences, and walls seven to eight feet high enclosed private courtyards and garden spaces; walls supported plants and were used

Grey wall, Wang Shi Yuan, Suzhou. In misty morning or evening lights this wall almost totally disappears: in the midday sun it seems as solid as a cliff.

to hold back the rush of water or the weight of earth. They were punctured to create 'leaking windows'; they sometimes went up and down the contours of a hill like a snake; they accentuated every change in direction or shift in height. Lovers were forever speaking, listening and, of course, climbing over walls – at least in plays and novels. Walls held up on their backs a tiny pitched roof of dark or coloured tiles. Reddish or greenish-brown walls could take on the colours of plants, moss or soil, or the patina of bronze. There is very little, in fact, that the Chinese did not do with this essentially prosaic element. And nothing in a garden was so fanciful as the doors and windows that were devised to break them open.

Such holes in a wall were very seldom rectangular. The main passageways into the garden quite often led through a circular shape, popularly called a 'moon gate', which was frequently without a door. Thus, the garden gate contrasted with the door of the dwelling as a circular or irregular shape versus a rectangle, as open versus closed. In China the circle is a symbol of heaven (just as the square symbolizes earth), and it is also an emblem of perfection. According to an ancient conceit, the moon gate, like the round frame of ancient bronze mirrors, provided the best possible setting for a view, and it is true that such gates focus down the eye, acting like the light-stop on a camera to intensify and concentrate all that is revealed beyond. This effect is caused not only by the round shape, but also partly by the dark edging which often surrounds it, or by the thickness of the wall itself when its edge is seen obliquely. In any case the dramatic intensity is undeniable, for the formal completeness of a circle contrasts sharply with the irregularity of the garden all around it. The moon gate also makes it necessary for the visitor to step into the garden over the curved bottom of the circle. He thus has to walk straight through the centre (unless he ducks), and for the same reason each visitor enters the garden singly, and the act of entrance is thus given special emphasis.

In addition to moon gates, the holes in garden walls are to be found in an abundant variety of shapes and metaphors, of which the commonest were the forms of petals, leaves, fans, and vases. The eccentric writer Li Yu used to cut out fan shapes from the walls of the boat he kept on the West Lake at Hangzhou: as he was rowed peacefully across the water, the moving scenery outside thus always appeared in a frame.

ABOVE, LEFT *Picture window, pavilion in Loquat Garden, Zhuo Zheng Yuan, Suzhou. The simple back wall of this airy* ting *provides just enough seclusion for those sitting inside, while the opening cut in it frames sunlit leaves and rocks as if in a painting.*

ABOVE *Moon gate within garden, Cang Lang Ting, Suzhou. A simple circle frames complicated patterns beyond, including a window shaped like a banana palm, with a real banana palm to its right.*

ABOVE *Oval window framing bamboo grove, modern park, Canton. Though most newly built parks are designed more for recreation and sport than contemplation, occasional summer houses still use traditional devices to highlight different scenes. Local guides point out that this is in harmony with Chairman Mao's teaching: Take what is good from the past to use in the present.*

RIGHT *Scholar playing* qin *in his peony garden, woodcut from Lin Qing's* Hong Xue Yin Yuan Tu Ji. *The rustic trellis overgrown with wisteria brings a 'flavour of woods and mountains' to this otherwise formal arrangement of rocks and peonies in raised stone beds. Note the use of a flat rock as a step up into the pavilion behind.*

LEFT   Eagle rock, with mirror image behind, Liu Yuan, Suzhou. Two monoliths set up in separate courtyards make the window in between seem like a mirror. This garden is full of such visual teases – inventive and delightful plays on what is 'real' and what is 'false'.

LEFT, BELOW   Ting with real mirror, Yi Yuan, Suzhou. The 'hole' in the back of this pavilion really is a mirror.

OPPOSITE:
ABOVE, LEFT   Design of cranes in a wooden lattice window, Yu Yuan, Shanghai.

ABOVE, RIGHT   Wisteria design in clay on wire. One of a group of four windows, Shi Zi Lin, Suzhou.

MIDDLE, LEFT   Round window, Suzhou.

MIDDLE, RIGHT   Square clay window with design of leaves, vase and shou (longevity) character, surrounded by swastika pattern, Suzhou.

BELOW, LEFT AND RIGHT   Windows on the Kunming lake, Yi He Yuan (Summer Palace), Peking. These little windows are double-glazed, with little paintings on the glass. On festival nights reflections from them lit up the lake surface below.

Windows in a garden could be even more fanciful in shape and subject matter than doors, especially in the late and often ostentatious gardens of the last dynasty. This was possible because their function was entirely decorative. Lacking the more fundamental purpose of a doorway, a window could be elaborated and filled in with all kinds of intricate detail. Ornamentation of this sort was often constructed of roof tiles composed into geometric patterns, or of clay and wire worked into scenes of birds and flowers, then baked and whitewashed.[5]

Perhaps no other feature in the garden lent itself to such excesses of naturalistic whimsy. The *Yuan Ye* again has a warning for the cultivated gentleman on this subject: he should be wary of the garden sculptor who would turn his peaceful retreat into a menagerie of fanciful beasts. 'In olden days it was common to have craftsmen decorate the walls with sculptured and engraved representations of birds, flowers, animals and fabulous beasts, which seem to be executed with

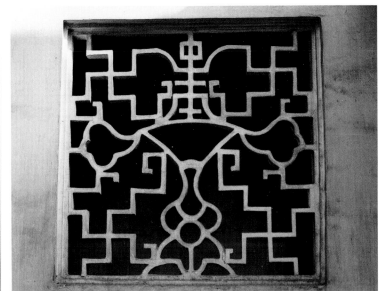

great skill, but such things look vulgar in the garden, and are not proper in front of the hall'.[6] Here the decorations have escaped from their window frames and begun to encroach on the blank white surface of the wall itself. But if the gentleman reader is still unconvinced, he is warned of the further perils of over-elaboration: 'Sparrows build their nests in [these ornaments] and grass grows over them as thick as creepers'.

Open Galleries and Latticework

Following the lines of the walls – and sometimes breaking free across the intervening spaces – the visitor will find that distinctive centipede of the Chinese garden, the roofed and open-sided gallery. Such galleries, called *lang* or *langfang*, wind up and down hills connecting pavilions and gateways, and at the same time dividing up the spaces like a screen. Tong Jun speaks of the

The Sui Yuan, Nanking, in the nineteenth century. Once owned by the writer Yuan Mei, this garden was famous for its covered galleries which made walking through it possible in all weathers. From this woodcut they seem to have linked only parts of the garden on the northern shore of the lake, but bridges and causeways allowed visitors to meander across the water itself. The garden was maintained thirty years after Yuan Mei's death as an act of filial piety by his sons. It was destroyed in the Pacific War. (From Lin Qing's Hong Xue Yin Yuan Tu Ji)

ABOVE   *Covered gallery or* lang, *former garden of Manchu prince, Hou Hai, Peking. Straight lines in this formal northern garden surround an irregular lake. The galleries in the southern gardens of Suzhou amble less directly through the garden, or zig-zag erratically through the pavilions.*

ABOVE, RIGHT   *Detail of balustrade, entrance courtyard, Shi Zi Lin, Suzhou.*

galleries as unifying the garden 'in one breath',[7] and the *Yuan Ye* insists that they 'should never be missing from any garden'. They served as a decorative frame for viewing and as a sheltered walkway between tea houses, halls, *ting* and *zhai*. The eighteenth-century writer Yuan Mei said that he could enjoy his garden (the Sui Yuan in Nanking), even in winter because of its long corridors that 'connect up with each other so that if there is thunder, and lightning with wind, there is still no need to stop walking'.[8]

Father Attiret, the eighteenth-century visitor, was surprised by such winding, open corridors in the Yuan Ming Yuan for 'that which is strange is that [they] never go in a straight line. They make a hundred detours, sometimes behind a clump of bushes, sometimes behind a rock, or sometimes round a lake; there is nothing so agreeable. In all this there is an enchanting, elevating impression of the countryside.'[9] Part of this enchantment comes from the unusual experience of being both inside and out at the same time, part from the decorative latticework that flows along on either side just above the eye, or below knee-level. This latticework is subdivided by the thin pillars which support the roof, a punctuation that gives each walk its own steady rhythm. Here and there along the way there are also resting places, where the *lang* widens a foot or two, or changes direction.

The patterns that make up the balustrades of these *lang* can dazzle the mind with their complexity, just like the whole garden labyrinth itself. They defy easy decoding, or rather they suggest an infinite number of geometric relations, depending on which elements are taken as figure or ground. The *Yuan Ye* classifies some sixty different patterns – favouring (as one might expect) the simple and coherent over the intricate. The forms which the author seems to prefer carry continuous geometric patterns, endlessly repeated so that they vibrate on the eye. Such designs are at once simple and fascinating, but not so demanding that they detract from the living patterns of shrubs and trees beyond them.[10]

In addition to the *lang* decorations, latticework was also used in the doors and windows of halls and pavilions. Most often it consists of rectangular patterns, for the obvious reason that this form packs most satisfactorily into architectural frames. However, these simple shapes were sometimes transformed by chopping certain squares to make a star or central cross, swastika, or cloud-spiral in the shape of an 'S' or 'U'. The swastika has been a favourite for the last three hundred years.

Placed within one sort of pattern it may symbolize prosperity and long life, since it relates to the Chinese word *wan* meaning 'ten thousand' – a synonym for 'forever' or 'innumerable'. Since its left-over parts formed 'L's and 'brush-handle' patterns, it was a marvellously ambiguous form, and one which was reasonably easy to build.

   The great importance of windows, even those which were very highly decorated, is underscored by a metaphor sometimes used to describe them: they are the 'eyes' of a house or pavilion. Without them the place is blind even dead; with them it comes to life.

Bridges and Paths

Closely linked to the *lang,* or open gallery, in function and decoration were the paths and bridges which continued the stroller's route, this time out in the open. Elegant bridges would pick up the patterns of the balustrades, sometimes reiterating them in marble, while paths of stone,

Zig-zag bridge to a 'heart of the lake ting', *garden of rest house for foreign guests, Guangdong province. Since evil spirits can only travel in straight lines, this pavilion on the water is well protected against unpleasant influences. The pond in this richly tropical climate is thickly covered in brilliant green waterweed, and contains a vociferous population of bullfrogs.*

RIGHT *Rock bridge, Suzhou.*

BELOW *Marble bridge, Yi He Yuan (Summer Palace), Peking.*

BELOW, RIGHT *Wisteria bridge, Liu Yuan, Suzhou.*

BOTTOM *Bridge with* ting, *Yi He Yuan (Summer Palace), Peking.*

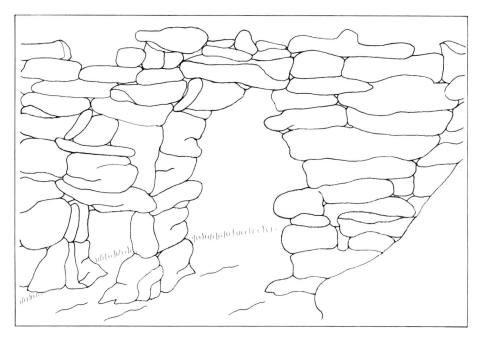

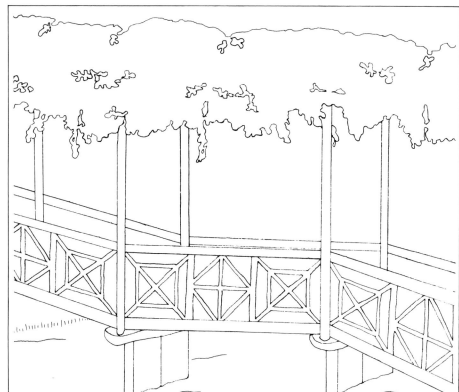

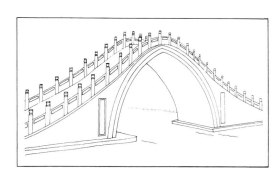

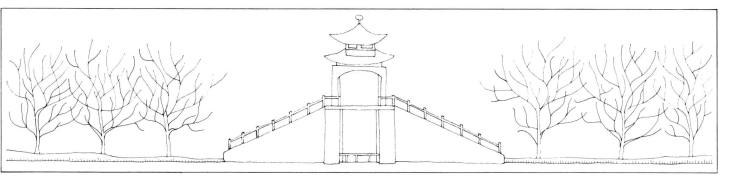

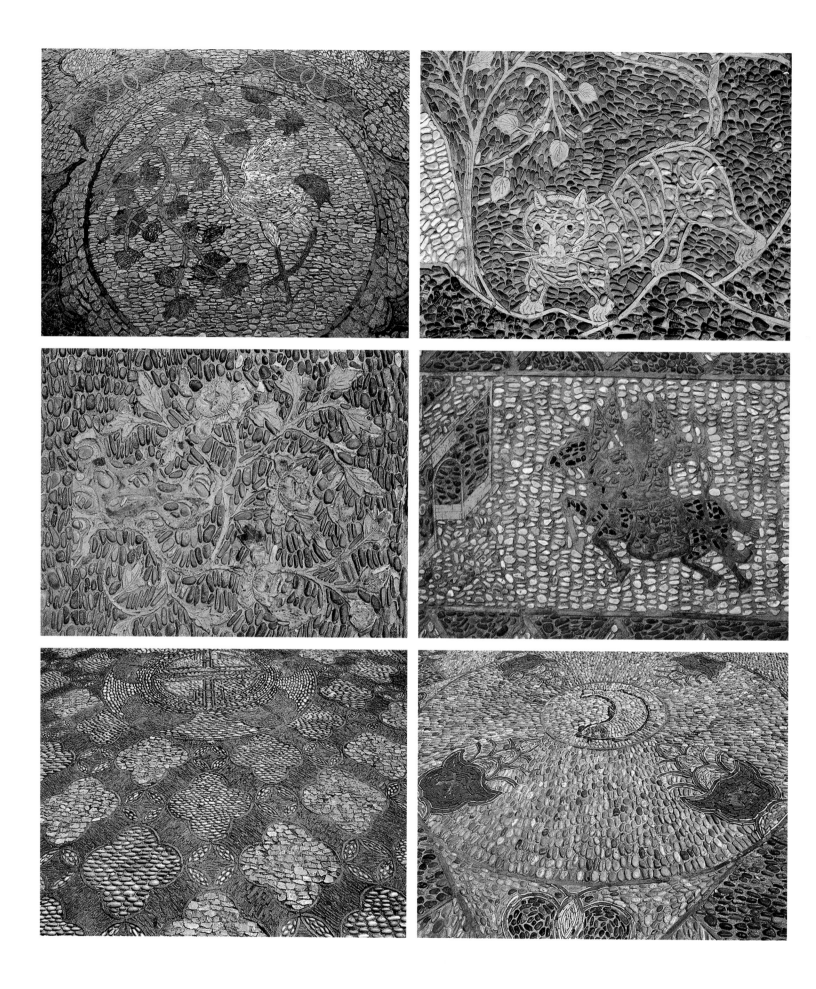

OPPOSITE PAGE:
TOP, LEFT *Courtyard pebble mosaic in Wang Shi Yuan, Suzhou, with crane and pine tree (longevity symbols), surrounded by bats (happiness).* RIGHT *Tiger mosaic in Yu Hua Yuan, Imperial City, Peking.*

MIDDLE, LEFT *Peony paving, Yu Hua Yuan.* RIGHT *Paving with general on horseback attacking a city gate, Imperial City.*

BOTTOM, LEFT *Paving in the Tui Si Yuan (Retreat to Think Garden), Tongli, Jiangsu: begonias surrounding a shou (longevity) character.* RIGHT *Paving in the Tui Si Yuan, with fish in centre, butterfly in foreground.*

THIS PAGE:
LEFT *Paving in the Imperial City, Peking, with figures from a historical drama*

ABOVE *Open doorway in the Zhuo Zheng Yuan, Suzhou. (See plan, page 183.) White walls undulate through this southern garden, dividing it into numerous compartments like the 'space cells' of a landscape scroll.*

157

LEFT, ABOVE *Internal garden wall, Zhuo Zheng Yuan, Suzhou. An open gallery follows the wall, branching out here to form the frame for a simple planting of bamboos and their shadows on the white surface.*

LEFT *Bamboo leaves and shadows, Zhuo Zheng Yuan, Suzhou.*

ABOVE *Pebble pictures from Yi He Yuan (Summer Palace), Peking.*

brick, or pebble mosaic would adopt forms such as the 'cracked ice' pattern. This transformation of repeated themes obviously helped to unify the variety of spaces and to harmonize the countless different incidents on a garden walk. A fine bridge might be one of these, its possibilities celebrated in white marble, with a halfmoon curve completed by its reflection below, and railings sculpted in the shape of lotuses or lions. But in simpler gardens plain slabs of stone might carry the visitor over a stream, or zig-zag to a pavilion in the centre of a lake.

The path mosaics formed further accents along the garden walk. Some patterns were simple and rustic, like the herringbone pattern made from bricks turned on their sides. Others, of different coloured pebbles, stone chips and roof tiles turned on edge allowed the designer to create all kinds of delicate geometries, and even representations of birds and flowers. The Summer Palace, Zhongshan Park and back garden of the Imperial City in Peking all contain charming pebble pictures: goldfish, historical scenes – even a motorcar and a man on a bicycle. And the Liu Yuan in Suzhou – that centre of scholarly elegance – is famous for its free patterns of flying cranes and Buddhist symbols.

Beyond all this, what distinguishes these paths from those found in the West is the constant change from one pattern to another, with a change in rhythm to announce a change in function. In a Chinese garden, even if the visitor walks around looking only at his feet, it is easy to see when the scene has shifted. Pebble mosaics define different spaces and emphasize the alternations in mood. One pebble mosaic may be formal and geometric, and the next one gentle and feminine, achieving the intimate mood of an embroidered carpet.

Lions and Couplets

While walking around a Chinese garden the visitor becomes conscious of small elements which call attention to the surrounding space. Among them may be a plate of dwarf trees set on a table, or a porcelain stool but, above all, an inscription over a doorway or on either side of an entrance. These markers are quite different from each other in function and form, yet they all serve to define a place and even spell out its feeling and use. Black tablets over a door were meant to impart a moral or supply a poetical description of the next space cell; while rampant lions, paired beside a gateway, were meant to welcome and impress a visitor.

Two stone lions are placed in the traditional style in front of the main gateway at the Yu Yuan, rearing back and baring their teeth, while inside the garden bronze lions twist their heads, look to the ground, raise a paw and snarl almost audibly. They are quite fierce and at the same time very beautiful and domestic, with swirling locks and dog-like features. In imperial gardens statues of other symbolic animals, especially deer and important reptiles such as the dragon and the tortoise (as well as large incense burners) are often placed in front of halls, adding formal

BELOW *Bronze ox, near beginning of the Seventeen Arch bridge, Yi He Yuan (Summer Palace), Peking. Engraved with calligraphy by the Qianlong Emperor, this fine beast specifically placates a difficult Dragon King who would otherwise cause floods in this area.*

ABOVE, LEFT *Bronze guardian lion, Imperial City, Peking.*

ABOVE, MIDDLE *Bronze tortoise, symbol of the North, Imperial City, Peking.*

ABOVE, RIGHT *Bronze elephant kneeling in submission, Imperial City, Peking.*

symmetry to the reception areas. Statues, however, are by no means as important to Chinese gardens as they are to the great gardens of the west. Indeed, the part played by classical goddesses and Greek temples in an eighteenth-century English park is in many ways played in China by the written word. The Chinese consciously evoked the richness of the past not by building in ancient styles or imitating antique statues, but by writing a carefully chosen name over a pavilion entrance in fine calligraphy. Such a name did far more than suggest a use. It put the experience of entering a pavilion into an historic and poetic context that engaged the mind of an educated man, just as the forms all around him played upon his senses.

A Chinese garden is as liberally scattered with such written signposts as a Chinese painting with colophons. In a painting the addition of calligraphy (various comments and short poems) and collectors' stamps enhances and develops the subject matter and, equally important, provides a certificate of authority and venerability which gives the object depth and meaning. Such colophons will often be plastered all over the sky of a landscape painting and may even obtrude, like the boundary walls of a greedy landowner, into the mountains and foothills. These stamps and calligraphic insertions are like picture captions in the Western sense, except that the Chinese allow them to serve as a counterweight in the composition. Most importantly however, they provide a running commentary, a palimpsest of meaning, an historical verdict – the best perception of the ages – without which a painting is incomplete. In China a garden is often compared to a piece of landscape art, and the inscriptions in a garden are perfectly analogous to the colophons in a painting.

Three kinds of writing are commonly found in gardens: names, inscribed couplets, and the appreciative poems of visitors, often written to commemorate particularly enjoyable days or elegant gatherings. It is said that rich merchants seeking literary respectability would give lavish garden parties in the hopes of perpetuating their fame in verse. The suggestion, of course, is that

LEFT *Standing stone tablet, Cang Lang Ting, Suzhou. This engraved stone records a visit of the Qianlong Emperor, who prided himself on his calligraphy. The Beihai Park in Peking contains many names of pavilions written in this Emperor's hand.*

LEFT, BELOW *Signboard over entrance, Peking. The strong calligraphic strokes look well against the brilliantly coloured patterns of the eaves.*

RIGHT *Tablets engraved with verse, Wang Shi Yuan, Suzhou. These poems record pleasant visits to the garden, and add to the new visitor's delight.*

they acquired their gardens more to this end than for any deep appreciation of 'hills and waters'. These *vers d'occasion* were engraved on to black polished stone tablets which were then sunk into the garden walls. The walls of the great Suzhou gardens are covered with such inscriptions, and indeed they were so common throughout China that two historic gardens were remarkable simply for having none at all. Here the visitor feels that something is missing, that the garden is in some sense unfinished. For the proper way to tend a garden was to increase its historical richness until finally it became another kind of labyrinth – a mental maze of scholarly interpretations and well-chosen metaphors.

As well as poetic commentaries, recording for posterity the insights and emotions of scholars who had visited a spot, a Chinese garden was also full of carefully chosen names, which were not merely designations but also acted as signposts pointing along a veritable highway of aesthetic, moral messages: 'The Gorge of Dripping Verdure', 'The Grotto of Secret Clouds', 'The Wind of Autumn over the Ocean of the World', 'The Centipede Bridge'. Ideally these signposts promoted a suitable mood, called up an apt metaphor, and made discreet reference to classical literature. Truly elegant names showed an understanding of traditional culture, with all its set associations, but were also creative choices adding new personal insights to old meanings, thereby extending the tradition and keeping it alive. Naming was thus a delightful yet very serious game.

A famous chapter in the *Dream of the Red Chamber* shows that the game could be tricky and amusing. The passage concerns the invention of inscriptions for the newly completed garden of the Jia family, a garden specially made for the visit of their daughter Yuanchun, recently elevated to become an Imperial Concubine. The naming of parts was entrusted to the eldest son Baoyu, as a kind of literary aptitude test to see if his schooling was adequate and if his creative ability was fresh without being too eccentric. Jia Zheng, the father, ruminates on the possibility of writing inscriptions before the Imperial Concubine arrives. It is interesting that, except for cousin Zhen who organized the actual building of the garden, the family party is now seeing it all for the first time:

LEFT *Corridor of windows and calligraphy, Suzhou. Portholes on the right frame views of the garden; latticework on the left screens the garden visitor from street life outside the garden wall – yet lets in light through opaque glass. Ahead a wall of engraved tablets is presented for him to muse on as he walks.*

'These inscriptions are going to be difficult,' [Jia Zheng] said eventually. 'By rights, of course, Her Grace should have the privilege of doing them herself; but she can scarcely be expected to make them up out of her head without having seen any of the views which they are to describe. On the other hand, if we wait until she has already visited the garden before asking her, half the pleasure of the visit will be lost. All those prospects and pavilions – even the rocks and trees and flowers will seem somehow incomplete without that touch of poetry which only the written word can lend a scene.'[11]

Eventually an attending literary gentleman thinks of a compromise: provisional names could be painted on temporary paper lanterns. The Imperial Concubine will then decide on the permanent names after she has inspected the garden. Both horizontal and vertical lanterns would be needed because apart from the name of each place – written horizontally over doors or under the interior eaves of pavilions – any fine garden would contain couplets written out on pairs of vertical panels to look like hanging scrolls of calligraphy.

Baoyu and his father, accompanied by the attendant *literati*, all perambulate the newly finished garden, remarking on its surprising aspects and thinking up plausible couplets. The first test of their literary ingenuity comes as they climb the mountain immediately inside the entrance, and kind a white rock whose surface is polished to a mirror smoothness. It is clear that this has been prepared for an inscription, so the literary gentlemen run through thirty names, most of which are literary clichés that are meant to be inferior to the one Baoyu will invent:

'Emerald Heights', said one.
'Embroidered Hill', said another.
Another proposed that they should call it 'Little Censer' after the famous Censer Peak in Kiangsi.
Another proposed 'Little Zhongnan'.

RIGHT *Stone engraved with the calligraphy of Marshal Zhu De, comrade of Chairman Mao, in the Orchid Garden (Lan Pu), Canton. The poem, entitled 'Visiting Yuexiu Park', reads: 'In Yuexiu Park is a grove of flowering trees; a hundred flowers bloom together, each competing for spring. Only the fragrance of the orchid is just right, the famous tribute of a time from the Five Rams' City [Canton].' Zhu De's signature and the date, 3 March 1961, follow.*

Thus, two familiar metaphors and two existing mountainscapes are playfully suggested in order to set up Baoyu for his little triumph. He makes good use of the opportunity, slightly mocking the pedantry of the literary men, but nevertheless pointing up the importance of venerability, tradition:

'I remember reading in some old book,' said Baoyu, 'that to recall old things is better than to invent new ones; and to re-cut an ancient text is better than to engrave a modern: we ought then to choose something old. But as this is not the garden's principal "mountain" or its chief vista, strictly speaking there is no justification for having an inscription here at all – unless it is to be something which implies that this is merely a first step towards more important things ahead. I suggest we should call it "Pathway to Mysteries" after the line in Chang Jian's poem about the mountain temple: "A path winds upwards to mysterious places." A name like that would be more distinguished.'

There was a chorus of praise from the literary gentlemen: 'Exactly right! Wonderful!'

The party proceeds through a tunnel of rock and thence to an 'artificial ravine' covered with trees and flowers. The description implies that the party has reached a vast expanse of undisturbed, idyllic scenery located in the mountains. There follows an exchange in which Baoyu names a pavilion 'Drenched Blossoms', which is an improvement (in realism) on the alternative 'Gushing Jade' and (in taste and appropriateness) on a line taken from the *Drunkard's Pavilion*.

Building in the Tian Yi Ge Library complex, founded in the sixteenth century, in Ningpo, Zhejiang province.

Having chosen the 'two words for the framed board on top', he now has to find the seven-word lines for the sides: "Baoyu glanced quickly round, seeking inspiration from the scene, and presently came up with the following couplet: "Three pole-thrust lengths of bankside willows green, one fragrant breath of bankside flowers sweet."" Everyone agrees this is imaginative and apt, just the right mixture of tradition and metaphor – at least their nods and praise seem to suggest this.

The company then proceed to a most secluded spot behind a whitewashed wall, where they find a *zhai* by a miniature stream, hidden among hundreds of bamboos – symbolic of the scholar's hut located far away in the crags and mists. The father rejects several names suggested by the gentlemen because they are too obvious, and then, with pride slightly concealed by censure, accepts his son's suggestion: 'The Phoenix Dance'. But he is less happy with his son's couplets for the *lian*, and like a tyrannical school teacher he implies that Baoyu's imagination is rather pedestrian.

The complicated naming ceremony – here the most significant act in garden making – goes on for an indeterminate time, taking in rock hills, scarlet balustraded bridges, herb garden, main reception hall, viewing terrace, nun's retreat, a clump of crab trees, concluding with a strange pavilion in a grove. Here everyone becomes subject to that magical confusion which is the essence of garden architecture. For it is a building without rooms (a contradiction in terms), made from 'corridors and alcoves and galleries . . . and partition walls'. The walls are built from wooden panelling with a medley of motifs, including 'bats in clouds, the "three friends of winter" – pine, plum and bamboo, little figures in landscapes, birds and flowers, scrollwork, antique bronze shapes, "good luck" and "long life" characters, and many others'. The confusion is further enhanced by false windows and doors:

> Jia Zheng, after taking no more than a couple of turns inside this confusing interior, was already lost. To the left of him was what appeared to be a door. To the right was a wall with a window in it. But on raising its portière he discovered the door to be a bookcase; and when looking back, he observed – what he had not noticed before – that the light coming in through the silk gauze of the window illuminated a passageway leading to an open doorway. But as he began walking towards it, a party of gentlemen similar to his own came advancing to meet him, and he realized that he was walking towards a large mirror. They were able to circumvent the mirror, but only to find an even more bewildering choice of doorways on the other side. . . . How very ingenious . . .'Follow me!' said Cousin Zhen, amused at the bewilderment of the others, who were now completely at sea as to their whereabouts. He led them round the foot of the 'mountain' – and there, miraculously, was a broad, flat path and the gate by which they had entered, rising majestically in front of them.

The Jia garden, a maze of seemingly infinite space and time, thus through its skilful layout manages to produce a sense of that surprise and enchantment which were taken as the most profound symbol of Immortality, and of the infinite Dao. It is easy to see why a Chinese garden might be thought of as a superior kind of religious experience.

ROCKS AND WATER

LEFT *Yu Yuan, Shanghai. The Yu Garden, originally created by a retired official in the sixteenth century for the enjoyment of his elderly father, has a number of notable decorative rocks.*

ABOVE *The tribute bearers (detail), attributed to the Tang dynasty painter Yan Liben (d. 673). Part of the tribute brought to a Han Emperor are some fantastic stalagmite-shaped rocks, worn into curious holes and hollows. A miniature bowl garden is also included in this exotic collection of offerings.*

Stone-worship and Petromania

Western visitors are often overwhelmed by the rocks they find in a Chinese garden. Grotesque and grey, tortuous and massive, they dominate the scene in much the same way that flower beds highlight an English garden. The Westerner is likely to be perplexed and curious about these sculptural shapes: why are there so many of them, why do heavy, contorted boulders in a dark colour take up so much room? They even seem slightly sinister – impressive, but alien to our taste for greenery, colour and lightness. There are several reasons for this peculiar Chinese preference, each of them stemming from a magical view of mountains and from the high metaphysical level on which rock worship is placed. 'Worship' is no exaggeration. The Chinese have loved and revered rocks almost in the way that we have admired and collected religious icons.

Stone-loving began in ancient times when mountains seemed to be imbued with supernatural power. In China five holy mountains came to symbolize the centre of the earth and its four corners. The Immortals were thought to live in the western Kunlun mountains or on magical islands in the eastern sea, and anyone who reached their abodes and conversed with them would perhaps gain the secret of eternal life. In the Han dynasty the Emperor Wudi tried to lure the Immortals down to his garden by building rocky islands that simulated their island dwellings; and Yuan Guanghan built a false mountain, the first of a long line. A description of a garden created in the sixth century by Zhang Lun, a minister for agriculture, brings out some of the features of such artificial mountains and hints at their hold upon the Chinese sensibility:

> He built up a mountain called Jingyang as if it were a work of Nature, with piled-up peaks and multiple ranges rising in steep succession, with deep ravines and caverns and gullies tortuously linked. So lofty were the forests, so gigantic the trees that sun and moon could not penetrate their shadowed obscurity; so luxuriant were the vines and creepers in their festooning as to control the passage of wind and mist. Here the [Taoist] adepts, the lovers of mountains and wilderness, might have roamed until they quite forgot to return to their heaven....[1]

Even large or strangely shaped boulders had long been felt to be powerful and were worshipped as local gods. They were thought to condense the potent wilderness of nature, and by sympathetic magic they too might confer on the garden owner a kind of immortality. In the Tang dynasty the appreciation of single rocks, now an aesthetic activity as much as a religious one, was elevated to connoisseurship, and was practised by cultivated gentlemen such as the political rivals Li Deyu and Niu Sengru, who both collected fine stones for their gardens. There was a particularly feverish outburst of rockworship, or petromania, in the eleventh and twelfth centuries. The Emperor Huizong's gigantic rock collection was financially crippling for the Empire. More modest in his ambitions was the celebrated twelfth-century stone-lover Mi Fu,

better known today for his fine calligraphy and poetry. He supposedly bowed deeply every day to a giant, sculptural rock in his garden and addressed it as 'elder brother'. Petromania broke out again six hundred years later and continued at a high pitch into the nineteenth century. When in the 1840s Lin Qing restored the Bai Shi Ting, or Pavilion for Obeisance to Rocks, in his Peking garden, he was continuing with greater intensity a tradition that was already two thousand years old.

The Chinese word for the landscape is *shan shui,* which literally means 'mountains and water'. This combination of elements inevitably evokes the Isles of the Immortals. It also suggests the fundamental opposition of *yang* and *yin,* of masculine strength and feminine moisture, and the aesthetically satisfying juxtaposition of rough with smooth, of still rock with flowing stream. The pairing takes on further meaning when we remember the ancient myth: rivers are the arteries of the earth's body, while mountains are its skeleton. These sets of opposites help to give some idea

ABOVE *The Bai Shi Pavilion, woodcut from the nineteenth-century illustrated journal of Lin Qing,* Hong Xue Yin Yuan Tu Ji. *This hall in the Ban Mu (or Half-acre) garden in Peking was originally designed by the writer Li Yu in the seventeenth century to house his rock collection. In the 1840s it was restored by the inspector-general Lin Qing, who is seen here admiring a fine Taihu stone from the porch. Within the pavilion other exhibits are visible, including a screen on the left made of a single piece of fine marble naturally veined with zig-zag lines that look like ranges of mountains.*

RIGHT *Rockery, Yu Hua Yuan (back garden of the Imperial City), Peking. Living trees grow from a huge rock pile, interspersed with petrified tree trunks.*

BELOW *Rock mountain, North-east corner, Shi Zi Lin, Suzhou. Close up, it is hard to find a focus for the eye which moves restlessly from rock to rock. Modern railings, on the left, have been added to the rock bridge spanning a deep gully.*

of the depth and intensity of the Chinese obsession with stones. Mountains, beautiful minerals, meteorites, and rocks battered into strange shapes, are not merely a superior form of sculpture (Henry Moore with rough edges), but have many layers of accumulated significance. At their best, pulled up from the bottom of Lake Tai where the most perfect rocks are formed, they are a concentration of the creative forces of the Dao.

The Li River at Yangshuo, near Guilin, Guangxi.

For a Chinese the feeling given by rocks in a garden is largely one of the picturesque and the wild. Climbing over or through them or pondering their suggestive shape, he would easily imagine being perched on a mountain wilderness, confronted by elemental forces. The experience is not calming – that effect is properly left to water – but confusing and unnerving. Such a reaction would be in agreement with ancient principles, for one of the instructions for rock design in the garden manual *Yuan Ye* is that rocks should 'appear wild, as in the Tiger Mountain in Suzhou'. 'Appear' is a key word, however, for although the garden rockery or the miniature mountain on a scholar's desk resembled the terrifying precipices of nature, the impression of wildness was nevertheless conventionalized. In choosing and polishing a rock, or building a rockery in a garden, the potent wildness of nature was stylized as well as concentrated. Moreover, however grand a rockery might have been with its ravines and forests, however 'craggy and precipitous' it might appear in description, it was still a miniaturization and not the real thing. For the Chinese, except for the celebrated Taoist hermits, the real wilderness was not entirely appealing. In miniature its qualities were easier to appreciate. And although the Westerner is quite right to feel momentarily upset and moved by the presence in the garden of so many tortured shapes, this disquiet is meant to be assimilated into a larger metaphysical peace. In a similar way the horror of crucifixion and the grotesque gargoyles in the Christian Church are part of an encompassing pattern that reconciles us to their immediate impact.

Rock Design

Both Li Yu and the *Yuan Ye* are emphatic that there are no exact rules for designing with rocks: everything depends on innate feeling. Nevertheless there certainly was a set of desired qualities, emphasized by tradition and reiterated by various writers. Before we enumerate these it is helpful to distinguish between the piled-up rock hill (or false mountain), the rocky shores of lakes and watercourses, and the single, strangely shaped rock. While the same qualities would be desired in all three, they were used and constructed differently. A false mountain would be made either from many large rocks piled up on top of each other, often forming hollowed caverns within, or from earth with rocks sticking out of it like the exposed elbows and teeth of some subterranean beast. In the second type, bushes and trees would grow out of the mound to strengthen it. Mountains entirely of rock and the rough stone-built banks of streams, on the other hand, were often quite bare of vegetation. To appear like mountains, rocks should not be set up in a row or a screen, and certainly not in an even, symmetrical sequence. Rather they should wind and undulate along as peaks do, giving a picturesque play of light and shadow that is carefully balanced. According to the *Yuan Ye*, one of the best sites for placing rocks is in the middle of a pond; here, equipped with a 'flying bridge' and formed into tunnels and grottoes, they become a fitting place to 'welcome the clouds and moon'.

The single grand rock would stand magnificently alone in front of a pavilion specially placed for viewing it, or be placed 'under a stately pine or be combined with wonderful flowers'. Smaller single rocks were placed on scholars' desks, and large, flat ones were used as garden seats and tables. But whatever the arrangement, in order to bring out the essence of nature (according to the *Yuan Ye*, 'the greatest of all artists'), it was necessary both to make a discerning selection and

LEFT *'Stone bamboo shoots', Imperial City, Peking. The designer of this corner has made a play on the name given to these needle-point stones by planting real bamboos around them. The wall behind is plastered and painted deep, rusty red. The contrast with the grey stones and shimmering green bamboo leaves is simple and beautiful.*

RIGHT *Table rock on a stand; an exceptionally beautiful small piece of natural rock which seems to spiral up from its narrow base in an elegant but dynamic curve.*

to heighten whatever effects nature had provided.

Li Yu, in his writing on rocks,[2] emphasizes the importance of the total composition over the ornamental brilliance of the parts. The art of building mountains is similar to essay writing, for in both cases the difficulty lies in constructing the whole piece. The rock builder can learn, he claims, from the essays of the great Tang and Song prose writers which stand out for their overall balance and inner *qi* (or forceful spirit); in such works the whole and the detail 'are the same'. One special problem is that a composition may flow in the details but fail to bear scrutiny from afar, and another is that individual rocks in a rockery are often proportionately too large. How, Li Yu asks, are we to achieve a unity of part and whole with very big rocks? He comes out on the side of miniature mountains made from a mixture of rocks and earth. If one makes a mud hill first and then places rocks on top of it, a subtle mixture of nature with art and a variety of mountain paths and scenery will result. The trees, shrubbery, rocks and earth will merge into a unified picture, whose individual elements are as much part of the whole as the brushstrokes of a landscape scroll – a comparison which is often invoked in writings about the art of composing with rocks.

Another analogy for placing rocks together was a social one: large and small stones should complement each other like a host and his guests, or a prince and his vassals. And there were also, of course, aesthetic considerations: one should match up the striations, harmonize the furrows and holes. A rock was valued for its bumps, wild hollows and scoops, or its anfractuous edges, or its spiky surface or craggy outline, or its horizontal graining; and such formal qualities naturally had to be accentuated either by contrast (placement against a white wall) or by reinforcement (inclusion within a medley of other grotesque shapes). The special qualities of

different types of rock, including the noise they made when struck, were well catalogued, and the residents of districts where fine specimens might be found often lived off rock-collecting. The *Yuan Ye* briefly describes fourteen different finding places in the provinces of Anhui and Jiangsu, but the author mentions that there are more he has not visited.

Li Yu mentions three formal aspects which should be sought in a fine rock and aimed for when building false mountains: *tou, shou* and *lou*. As usual in Chinese, all three terms are ambiguous. The first, literally meaning 'go through', seems to imply that one should be able to walk through a passage or cavern from one point to another, either physically or in the imagination. But its overtones of 'holey', 'vulnerable', 'transparent', 'hardly there', are perhaps even more important. These connotations connect up with the second desideratum, *shou*, meaning 'thin'. The word suggests the delicate, feminine fragility that could be sought even in rocks, which usually represented the *yang* (or 'masculine') aspect of a landscape, or the emaciation of the ascetic hermit. Several blue and green paintings of the seventeenth century show just how delicate rocks could be. Some were almost like a filigree of mica. The term *shou* also suggests 'without visible support', 'upright in isolation', and indeed these rocks often seem to balance on a point. The *Yuan Ye* advises that single rocks should float like a cloud, with the heavier, broader weight at the top, and the narrower, thinner part at the base. Then they would 'fly and dance', or have the wild, fearful aspect of 'overhanging cliffs'.

The third quality, *lou,* literally means 'leak' or 'drip', and in this connection perhaps refers to an upright opening, a hole perpendicular to the surface, the desired effect being a series of such small holes open to all sides. An example of fine *lou* is the Exquisite Jade stone in Shanghai whose 'leaks' were also called 'eyes'. Opening on all four sides, they could see everywhere and thus suggested wisdom.

When one reads the various descriptions of stone types in the *Yuan Ye*, a picture of the stone connoisseur begins to emerge: a very versatile gentleman – part aesthete, part mystic and part char-woman with her scouring brush (or rather char-woman's supervisor). Many rocks had to be cleaned up and sometimes improved slightly, as for instance:

> Lingbi stones from near Suzhou . . . are found buried in the earth under layers tens of feet in thickness. The reddish clay is stuck so fast that they must be scraped with metal knives. Not until after the scrapings does the colour of the stones appear, and after this they must be polished and ground with brushes of steel wire or bamboo and pulverized porcelain. If one strikes them, they give forth a metallic sound. On the underside the earth has penetrated so deeply that it cannot be removed. These stones remind one of various objects; some of them are like mountain peaks, sharply cut and riddled with holes. But the hollows seldom form a beautiful pattern; they have to be chiselled and polished for their beauty to emerge.[3]

We imagine our stone collector labouring away on his love – observing the furrows and holes, the colour and texture of surface; giving the specimen under investigation a ping to see what music it makes; wondering if it looks like a lion or a Buddha, or what cloud formations and landscapes it resembles; perhaps, if he is a Taoist and academically inclined, remarking on the veins of mineral and how they might have some strange magical or at least medicinal purpose; and even wondering, if it is not already coloured, whether he: might not smoke or dye it to bring out a brilliant polychromy. The stone is coated with a layer of red earth? It must be a Kunshan stone. Scrape out its hollows and strike it. If there is no sound, then the origin is proved and the future of the stone resolved: it must be used in a miniature bowl garden, not as part of a large rockery.

Du Wan's catalogue of stones, *Yunlin Shipu* (Stone Catalogue of the Cloudy Forest), written around 1125, provides us with a detailed description of Taihu stones (which came from the bottom of Lake Tai):

> Huge specimens, up to fifty feet high, with a colour range from white through pale blue to blue-black, their surfaces textured in net-like relief, are hauled out of the lake. The most desirable have tortuous, rugged contours, and abundant hollows. Small surface cavities are called 'arbalest pellet nests'.[4]

These cavities were made by the hammering of waves, which in stormy weather ground smaller, harder pebbles into the porous limestone of which the rocks were made. Although genuine Taihu stones lost value if subsequently improved by human hands, a local industry developed which encouraged the natural weathering process by dropping suitable specimens of limestone into the lake. The bigger the holes, the more valuable and delicate the rocks – and the greater the risk of their breaking apart. Large specimens became increasingly rare and fetched enormous prices. Connoisseurs were constantly warned against forgeries.

Lake Tai stones fell into the category of the 'baroque', as defined in Du Wan's catalogue. A 'baroque' stone was one that was 'odd', 'singular' or 'fantastic'. Opposed to this class was that of the 'primitive', describing rocks that were valued for their mimetic quality – for natural interior colouring resembling clouds, mountains or forests, or for looking like animals, deities or dragons.

ABOVE *The Cloud Crowned Peak (Guan Yun Feng), a highly esteemed rock in the Liu Yuan, Suzhou.*

LEFT *Stone lion, courtyard of the Shi Zi Lin, Suzhou. The name of this garden means 'Lion Grove', and it includes many rocks which resemble the heads and bodies of lions – some more ambiguous than this merrily prancing creature. In fact no lions are indigenous to China, so it is an almost mythical beast which artists could portray from the imagination, unhindered by any direct experience with the animal in reality.*

RIGHT, ABOVE *Ting beside a causeway, Li Yuan, Wuxi. The causeway and trees separate the larger lake from a smaller pond. Without them the mountains and water would be very much less interesting.*

RIGHT *The historian Sima Guang in his home-made pavilion, painting by Qiu Ying, c. 1510–51 (see pages 96–7). Although mountains and water are essential elements of the ideal Chinese landscape, without a ting, says one authority, it can hardly be called a garden, (see pages 132–3).*

In the Shi Zi Lin (or Lion Grove) in Suzhou the tumbling piles often resemble lions. The rockery was apparently laid out in the year 1342 by a famous monk, Weize, who wanted to remind himself of his former habitation – the Lion Rock on Tianmu Mountain. The Lion Grove is thus a double mimesis, a substitute for a real mountain that in turn looks like a lion. As the visitor walks through and around the rock piles here, bushy manes, mouths, tails and paws form and re-form among the stones and hollows, between old, gnarled pines, ancient petrified trees and needle points of limestone stalagmites (known as 'stone bamboo shoots'). All these elements vie with each other in a contest of grotesquerie so that, as with certain Op Art and Islamic patterns, the eye is continuously moved on in a dazzle of light and shadow, solid and void. When one adds to this the profusion of suggestive shapes (a veritable Rorschach test of recognizing species), it is easy to see how the viewer's sense of relative size is confused, so that space, and therefore time, are suspended in the garden. Wandering among the contorted peaks reflected luminously in the central lake, it is not difficult to image that when the guides and visitors have departed for the day, Immortals – unspeakably ugly, perhaps drunk, but no doubt inspiring in their eerie paradise – will materialize here on the backs of storks.

Island of Small Seas, West Lake, Hangzhou. The island is composed of clover-shaped causeways separating four lotus pools from the lake beyond. Tradition suggests that it was built by the poet Su Dongpo, governor of Hangzhou from 1089 to 1091, but part of the island garden's fascination comes from much later additions – an apparently random scattering of pavilions and zig-zag bridges which delicately break up the symmetry of the island itself.

Water

'Where there are mountains,' goes an old saying, 'there is bound to be water in the same place.' When Emperor Ming of Wei (AD 222–240) found his park contained everything except water, he had foot-pedalled water wheels installed to bring it in artificially: only then was the park

Part of the rockery, Shi Zi Lin, Suzhou. This garden has changed greatly from its original form in the sixteenth century (see page 117) and now includes the largest 'false mountain' in Suzhou. Inside it is a labyrinth of caverns, chasms, gulleys and tunnels which twist through the rocks. Outside the different shapes resemble animals, faces, figures.

considered complete. To the Chinese water is not just physically beautiful in a garden, but is absolutely necessary to balance the mountains, and so to represent the totality of Nature in perfect harmony. We have mentioned the complementary aspects of *yin* and *yang*, and this formula was quite explicitly applied to water and mountains. The mountain here represents the masculine *yang*: upright, bright, hard and bony. By contrast water in its *yin* aspects is receptive, yielding, wet and dark. A famous saying in the *Analects* of Confucius, however, reverses the *yin-yang* relationship of mountains and water; here, water becomes the active principle, expressing itself in swiftly moving torrents, while mountains become the passive, reflective principle because they remain fixed and motionless. Thus, runs the passage, 'the wise man delights in water, the good man delights in mountains. For the wise move but the good stay still. The wise are happy but the good secure.'[5] Mountains and water, like all other things, always contain within themselves some aspects of their mirror opposites.

It is usually noted that the traditional gardens of China make no use of fountains, since these force water to do things against its own nature. On the whole this is true, but in imperial gardens there has been quite a lively tradition of fanciful water-toys – known as 'hydraulic elegancies' or *shui shi* – which momentarily entertained various Emperors. In the old Song capital of Kaifeng in the twelfth century there were two white lions with Buddhas on their backs who spouted streams of clear water from their fingertips. Later, the last Emperor of the Yuan dynasty (1280–1368) had a series of fantastic fountains built for him incorporating balls that danced on jets of water, tiger robots, dragons exhaling scented mist, and boats of whirling mechanical

LEFT *Summer pavilion, Pond of the Four Seasons, Li Yuan, Wuxi, with Spring pavilion to the left. A 1957 addition to this merchant's garden of the early twentieth century. The pond has been literally 'borrowed' from the lake by means of the causeway on which stands this pavilion, but visually the viewer 'borrows' the lake stretching away behind the pavilion, and it becomes part of the garden.*

ABOVE *Dragon-headed spring, Imperial City, Peking.*

figures similar to those of the Emperor Sui Yangdi (whose extravagances we have already noted). A last essay into the 'unnatural' use of water was made under the Qianlong Emperor's auspices when, for the 'European' section of the Yuan Ming Yuan, he commissioned the Jesuit Father Benoit to build a series of life-sized animals who spouted water on the hour to tell the time.

Although private gardens did not attempt such shows – which were regarded as somewhat barbarian novelties – they did make use of dragon-headed springs, from whose scaly mouths fresh water bubbled into placid pools. Also, gardeners liked to include, whenever possible, the different sounds of water gurgling over pebbles and rocks, dashing down steep gullies and trickling gently from the ends of bamboo pipes. Such effects were found naturally in the mountain sites where poets like Xie Lingyun and Bai Juyi had built their simple retreats. Nevertheless the most characteristic use of water in Chinese gardens concentrates on its *yin* aspects – on water as peaceful opposition and balance to mountain scenery.

A twentieth-century writer, Yang Hongxun, remarks that it is water above all that creates the mild, congenial and yet lively atmosphere which, since it best revives the spirit, is the ultimate purpose of a garden. Nothing else produces quite the same serene effect as a great lake, where the hazy water merges imperceptibly into watery mountains and clouds. In such places, the visitor 'stands alone in opposition to space', and literally is 'divinely delighted with its pure expanse'.[6] However, truly boundless horizons are not for gardens. Just as the 'borrowed' landscape (*jie jing*) in the mountains, although basically untouched, needs buildings and paths to highlight it, so the shores of great lakes need little islands or promontories with pavilions on them, to enhance the distance and bring out the experience of being in a picture with 'ten *li* of smokey rain and sea and sky all in one colour'.[7] Even on the edge of the great Lake Tai then, a little labour added to nature makes all the difference, while the smaller lakes of China's most famous beauty spots are essentially all manmade. The West Lake at Hangzhou, the Kunming Lake of the Summer Palace, and the newly made Seven Star Crag Lakes in Guangdong, were all more or less created out of ancient marsh lands, but so skilfully dredged and so carefully dammed that they look more natural than nature.

The poet Su Dongpo (Su Shi), once personified West Lake as a young girl, lovely in her everyday clothes, but still more beautiful dressed up. This lake, like many others, had been extended and then embellished by islands, pavilions, causeways and bridges, which brought out all the delicate artistry of irregular shorelines and rocky banks – and were reflected, with their background of hills, on the watery surface.

In reflection, semi-circular bridges complete themselves, producing the ideal round shape – a symbol in China of the moon and of perfection. The mirroring water immensely increases the beauty of the pavilions and hills, while the buildings and their reflection heighten the magical effect of the water's luminous surface. For these aesthetic reasons as well as for the harmonious balance of *yin* and *yang*, the garden manual *Yuan Ye* decided that the best sites of all for a garden are on the edges of lakes with a view of mountains beyond. Whatever the site, however, it always had to be selected with a view to the water supply, and no foundations could be dug before the

sources and flow of water had been fully investigated. With earth left over from digging streams
and dredging pools the gardener would make hills to augment the natural terrain, so that even
in the countryside 'digging ponds and piling mountains' was a common Chinese phrase for
making a garden.

In city gardens, of course, this was a wholly accurate description. Here, much smaller areas of
water had to represent the great lake views of the ideal sites. Little pools dug out near *ting* or
beside the main reception halls had to evoke depth and distance as well as reflect the buildings,
and streams a few yards long had to seem as if they had flowed into the garden from the distant
hills. Hence the many meanders and wanderings of the garden's pools and rivulets, their sudden
re-appearance round the next corner, and that characteristically Chinese trick, the whitewashed
wall that opens below letting the stream pass through but blocking the visitor's progress.

Often in city gardens the water is broken up into many small, scattered areas instead of flowing
into one larger lake, on the Chinese garden principle of 'divide and multiply'. The Zhuo Zheng
Yuan (or Artless Administrator's Garden) in Suzhou is composed of as much water as land, and
is famous for its complex, interconnecting pools. In fact it is almost a water-labyrinth – and much
more difficult to unravel when one is actually in the garden than it is in plan. In the central part
of this garden the islands in the lake are so large it is impossible to see round them, and in the
imagination the water may seem to flow away beyond them into infinity. Architecture divides
and complements every different part with a network of covered bridges, pavilions on stilts,
terrace-promontories and walls, so that each area of water seems quite unlike the others. The
bright sparkle of the larger pools is contrasted with the cool and sombre pond behind the main
reception hall (F on plan), where the banks are built up steeply into rocky cliffs, and huge dark
magnolias overshadow the surface. Beyond this the main water area flows on and around, under
a room built over stilts, and into a peaceful backwater (J).

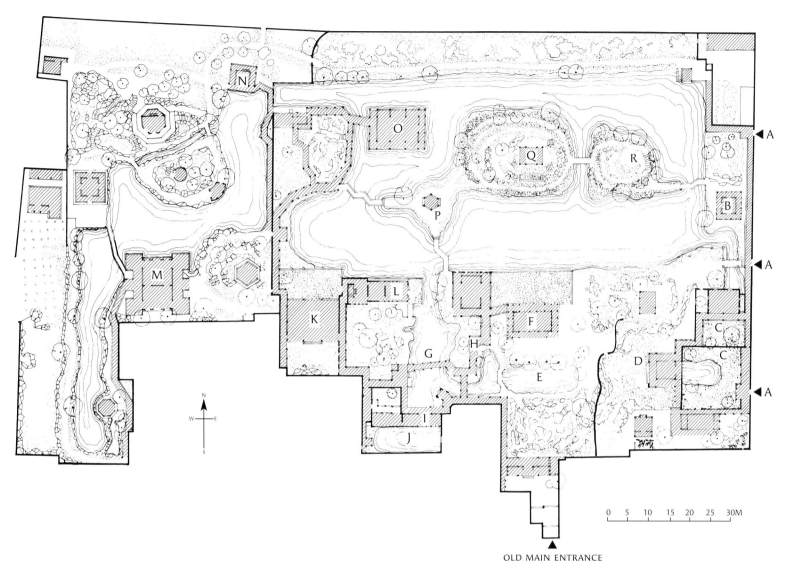

OLD MAIN ENTRANCE

Zhuo Zheng Yuan, Suzhou

─── – walls

▬ ▬ – doors

▭▭▭ – walls with windows

═══ – walls below knee height

▭ᆸ▭ – low walls with pillars supporting roofs

– false mountains, rock piles and hills

– trees and shrubs

– pebble pavings

– dwarf trees and flowers in pots

A   Entrances from eastern part of garden

B   'Quiet retreat among bamboo and wutong trees'

C   Small enclosed courtyard

D   Loquat garden

E   Deep pool

F   'Hall of Distant Fragrance'

C   'Little Flying Rainbow Bridge'

H   Open galleries

I   'Clarity and Magnificence Pavilion'

J   'Pure Ambition and Distant Thoughts"

K   'Flowers of the Pen Hall'

L   Tea house marble boat 'Travelling at Random'

M   Mandarin Duck Hall, with blue glass windows

N   Hall with mementoes of Wen Zhengming

O   '*Lou* for viewing the mountains'

P   'Lotus Wind on all sides' *ting*

Q   'Fragrant Snow, Beautiful Clouds' *ting*

R   'Persuading One to Farm Diligently' *ting*

The water in these curving lakes is by no means always clean and brilliant The Chinese are not worried by murky water and, characteristically making a virtue of necessity, describe opaque and milky streams as pearl-like. Green water is 'cloudy like jade' and much appreciated when a light breeze blows across its surface turning it into 'jade waves and clear wrinkles'.

Other typical qualities that were cultivated focused on the cooling or refreshing aspects of water. Naturally these were sought in the long months of summer heat, which both in the south and in Peking can mean temperatures in the nineties for weeks at a time. Pavilions in the Beihai Gardens in Peking include the Pavilion of Diffused Coolness, the Water Cloud Kiosk and the Hall of Lapping Waves, as if the names themselves could summon up a breeze across the lakes. All kinds of effects were used to highlight coolness in summer, even sound effects (which water heightens): the rustling of densely packed lotuses in a breeze; the pattering of rain on their umbrella-like leaves.

Indeed, the pleasures of water were found as much in its side effects as in the placid beauty of its reflections. A scholar might spend all afternoon in the summer heat in a pavilion set in the middle of his garden pond. Here he would lose his sense of individual being looking down at the fish and aquatic grasses softly waving beneath the surface, and breathe in the scent of lotus flowers and pine trees on the breeze. Perhaps he might dangle a fishing rod over the pool, not really wanting to catch anything. And later he might lean over and, like so many others before him, hold the reflection of the moon in the palm of his hand.

However, in elaborate gardens (especially in the north) streams and lakes did not always lend themselves to such informal pleasures. In the Yuan Ming Yuan and the great Peking gardens of Manchu aristocrats, water was sometimes shaped into formal patterns, such as swastikas, by enclosing it in crisply chiselled stonework with pavilions built alongside following the same design. Such highly stylized effects were used in deliberate contrast to the softer meanders of waterways in other parts of the gardens. The great Tang poet Bai Juyi perhaps had a similar – though less elaborate – contrast in mind when he built a 'square pool' among the pines of his mountain estate on Lu Shan, but the stiff angularity of the Manchu designs remains very far in spirit from the naturalistic effects sought by scholar-gardeners in the south.

Metaphysical Reflections

We have mentioned the aesthetic qualities which were sought in water. All these also had a much deeper metaphysical or magical meaning which, however diffuse or attenuated in later centuries, still played an underground role. When a literary gentleman went to his garden pavilion of an evening 'to watch the moon washing its soul', he was calling, however subconsciously, on a system of magical beliefs and images that went back to very ancient origins. For the reflective man, the beauty of water in a garden would take on a spiritual significance since water was one of the Taoists' favourite symbols. 'The highest good,' says the *Dao De Jing*, is like that of water. The goodness of water is that it benefits the ten thousand creatures, yet itself does not wrangle but is content with the places all men disdain. It is this that makes water so near to the Dao.'[8] It could 'wash' away evil and so symbolize benevolence, quietly find its own level, thus suggesting perfect rectitude, and reflect like a mirror without egotistical distortions.

Since water had these properties, it is not suprising that a Taoist of the fifth century BC, Guanzi, put it forward as the original element constituting everything that exists – an idea which recalls the ancient myth of Pan Gu, whose body became the earth and whose blood flows in the rivers and streams: 'Water is the blood and breath of the earth, flowing and communicating as if in sinews and veins'.[9] This poetic conceit is later applied to gardens: streams are the blood that gives life to the scenery and water the animating spirit which provides delight and refreshment. Even Confucius could not avoid praising the harmonious aspects of water, and like the Taoists he recommended it for its moral properties. 'In a water level', he said, 'the water is in a most perfect state of repose. Let that be your model. The water remains quietly within, and does not overflow. It is from the cultivation of such harmony that virtue results.'[10]

In ancient China, when so much depended on the yearly rain cycle, it is not surprising to kind water infused with the power of immortality. Many stories relate the good fortune of people who actually managed to meet the Immortals, the *xian*, and drink with them from their tiny but inexhaustible containers. If the elixir had its beneficient effect, they would then shrink in size to enter through the narrow throat of a double gourd so that they could visit the underground palaces of the Immortals. Immortality was spoken of as 'eternal life beneath the waves', and a grotto might give access to the subterranean realms of the *xian* and therefore to heaven. Not surprisingly, we find these meanings incorporated directly into garden buildings, and rocky caverns built by the side of lakes and streams.

The celebrated kings of ancient legend who searched grottoes and mountains for the immortalizing fluid sucked up the nutritive moisture that seeped from the ends of stalactites and from the rocky walls of deep caverns. Stalactites themselves became potent through this association and were represented on ancient Chinese bells as protuberant knobs. The king's ritual drinking was known as 'sucking on the teats of the Celestial Bell'[11] – a connection of water with women which provides all sorts of surprises and opportunities for Freudian speculation.

Since women were the *yin* aspect, they should be properly harmonized and balanced with the *yang*. Catastrophes followed if the scale tipped too far to either side, and the balance had to be restored by imperial intervention. Thus, most spectacularly, when the country was devastated by an appalling flood in AD 813, the Emperor immediately expelled 'two hundred wagon-loads of superfluous women from his palace', and the overflowing rivers dutifully returned to their beds. In fact women 'represented metaphysical water in human form' (as Edward Schafer puts it). 'They appeared in mythology and in literature as visible forms of the moist soil and the water-courses that make it wet.'[12]

In very ancient times Chinese water goddesses seem to have been inhuman, scaly creatures, rather like the serpents and mermaids of other early cultures. In China they were coupled with the benevolent, rain-bringing dragon (so familiar in Chinese art) to represent two different aspects of water. Schafer explains that the serpentine ladies were 'the repositories of moisture – the cool, receptive loam, or the lake or marsh; [while by contrast] the virile dragons were the active, falling rain'.[13] As the centuries wore on, the dragons became increasingly male in nature and the ladies less scaly and infinitely more enticing. Thus, in medieval times, the guardian spirit of a part of the Yangtze gorges appeared to men luminously veiled in drifting rain or morning mist. Tang dynasty poets wrote romantic verses by her temple, remembering the tale of an ancient king whose reign prospered after an amorous assignation with this lady. She was known as the Rainbow Woman, and both the rainbow and mist became associated with her supernatural powers of fertility.

If lakes and waterways were appreciated in the morning mists, they were equally celebrated by moonlight. In some places the water goddess was also the moon goddess. Even when the deities were separate the moon was associated with water, or ice; and because it was linked to the aqueous nymphs of the lake by the symbol of a pearl (thought to be composed of the moon's solidified essence) it was also the celestial image of woman. Such associations might have travelled through the mind of a literary gentleman as he sat in the lakeside pavilion watching the moon's reflection, for the *Yuan Ye* transforms them into ideas for garden design: 'The moon shines through the willow trees by the pond when it "washes its soul" in the clear water' (one might have entered the garden through a moon gate on the way to see this). The subterranean land of the Immortals might also be suggested, for 'the mirror of the pond reflects the shadows; here is opened an entrance to the mermaid's palace'.

Buddhist rock carvings, Ling Yin Temple, Hangzhou.

Even on rainy nights, however, when there was no moon, or on days of drizzle and mist, the pavilion on the water remained delightful to the Chinese. For like the Japanese and the English they have made a virtue of inclement weather, and the wispy vapours which arise from lakes and rivers have always remained full of romantic promise, even after the Chinese had ceased to believe in the river goddesses themselves. Certain lakes and gardens are famous for their beauty under somewhat damp (or 'feminine') conditions. The Kangxi Emperor of the Qing dynasty, for instance, made a special trip to a celebrated island garden known as the Tower of the Misty Rain (Yan Yu Lou). When he arrived the sun shone brilliantly, the waters sparkled and the outlines of the isle shimmered in the heat. It was a perfect day; and the Emperor was deeply disappointed. Had he arrived at the garden on a proper day of drizzling rain and shifting fog, this emperor would no doubt have composed a few verses to the local water goddess. But his inspiration would have been poetic and not supernatural. Even by the tenth century, the power of magic and the belief in water goddesses had been curtailed by the educated classes. Confucian propriety and rationalism had checked such thoughts, or deflected them in an aesthetic direction. On the poetic level there is no contradiction between the various meanings that water possessed – the ultimate essence, the blood of the earth, the undistorting mirror, the Immortals' drink, the place for mermaids, and woman's essence, which was summarized in the moon and crystallized in a pearl. Water had the power to conjure up all these related associations for a garden visitor inclined to receive them, and such a chain of ideas would often play a role, even if sometimes only a subconscious role, in the design and appreciation of gardens.

At the very least, it is water that brings life and movement to the rocks and summer houses. Slowly it pulses through the winding streams, mirroring the changing sky, and suggesting another world beneath its half-reflecting surface. When a garden loses its water, when the beds dry up as they do when neglected or during drought, then the aesthetic life is gone, not to mention the actual life of bamboos and lotuses. It is not surprising that some of the greatest Chinese gardens are built on lakes, and that the most famous gardens were made in the two southern cities in which water runs everywhere through rivers and canals. The saying, 'Heaven above, Hangzhou and Suzhou below', partly owes its inspiration to the prominence of water in both cities.

West Lake

Marco Polo, when he visited Kinsai (Hangzhou) in the thirteenth century, thought it the most splendid city in the world, and that a voyage on its West Lake offered 'more refreshment and delectation than any other experience on earth'.[14] Certainly no city in China was more suited to a life of pleasure and the cultivation of both gardens and art. Here the great Academy of Southern Song had found a proper setting for its celebration of mood paintings. Even today the rolling hills and mild warm climate, the mists veiling the landscape, and the vast expanse of peaceful water, all still provide one of the most famous symbols of beautiful landscape in China. Pencil-thin causeways divide the lake, thus allowing double reflections as one walks along them. There is no edge between tree and water, or land and water – no vast retaining walls or borders such as the eye is used to – so that the causeways and their trees seem to be floating

miraculously on the still, mirror-like surface.

Many of the old private gardens around West Lake retain a memory in their names of the prominence of water: The Villa of the Golden Stream, Residence of Water Bamboo, Rippling Water Garden, and so on. None of them is at present open to visitors, while many of the monasteries and Buddhist temples that once surrounded the lake have not survived. Fortunately, however, the most impressive remains, the Ling Yin Si, which was founded in the ninth century. Located in the southern hills, it is approached through a dense grove of bamboos which used to provide shade for streams of pilgrims. There are airy pavilions whose roofs are embowered in the trees, a stone pagoda, a high-arched bridge and a marvellous series of moss-covered Buddhas chiselled into the side of the mountain. In the ninth century the southern hills were full of such monasteries, and their roofs could be seen from the lake, lifted above the tree line, a series of swooping eaves and thin spires. The poet Bai Juyi, appointed Governor of Hangzhou in 822, visited these monasteries and wrote of their charms. His poetic descriptions have fascinated generations and captured the singular peacefulness of Hangzhou. In 823 he wrote the *Record of the Cold Fountain Arbour,* an account of a small building surrounded by water that lay close to the Ling Yin monastery:

In spring I love the smell of grass, the filling out of the green upon the trees, which sooth and purify the spirit and exhilarate the humours of the blood. On summer nights I love the trickling of the fountain, the chill of the breeze, which wash care away and dissolve the fumes of wine. Here the mountain trees are my roof, the rocky cliffs my screen. Clouds rise from the rafters of the shrine; the water is level with its steps. As you sit and enjoy this scene you may wash your feet without rising from your couch; while you lie in intimate converse with it, you may dangle your fish-hook with your hand still on the pillow. Nay, more! So clean and clear is the flowing stream, so pure and fresh, so soft and slippery that whether you be layman or monk, the mere sight of it will take the dust from eye and ear, the grime from heart and tongue, without the need for washing or rinsing.[15]

Water, even if no longer the ultimate essence of life, has here become its poetic equivalent in a way that is still appreciated by the Chinese in gardens today.

FLOWERS, TREES AND HERBS

LEFT *Leaves of* Ginkgo biloba.

ABOVE *Fa-tee (Hua tian) nursery garden, Canton.*

The love of flowers is an ancient passion among the Chinese, and this is hardly surprising; among all the flowery regions of the world it would be hard to kind one to match China in the richness and diversity of its natural flora. Today one cannot walk through a garden in England or North America without funding species of azalea and rhododendron, of rugosa and Banksia roses, lilacs, camellias, daphnes, lilies, primulas, large-flowered clematis and countless others which either first made their appearance in China or have since been developed in the West from those that did.

Without the Chinese we would not have garden chrysanthemums, nor the huge showy blooms of tree peonies which Marco Polo once described – to everyone's derision – as 'roses, big as cabbages'. Peaches (called *Prunus persica*) first grew in China, and from there, during the Han dynasty, found their way to Persia via the Silk route across Asia. Certainly the sweet orange was first discovered in the orchards of south China by the Portuguese, who took it back to their Indian settlements, and thence to Portugal in the sixteenth century. After their formation in the early 1600s, both the English and Dutch East India Companies did a regular traffic in plants from the nursery gardens around Canton. From these sources the tree peony finally arrived in England in 1787, as well as the continuously blooming China roses which were to have such a profound effect on all modern blooms. This gathering of flora was fairly haphazard, however, until 1842, when the plant collector Robert Fortune arrived in China on his first field trip for the Royal Horticultural Society in London. During several visits over twenty years he covered every nursery garden from the famous Fa-tee (*hua di*) gardens of Canton to the nurseries of Peking. He also occasionally visited private gardens of rich merchants, and even of scholars, noting their odd rocks and 'fairy-like' scenes with a certain begrudging admiration.

Starting off in the south, he sent back a new weigela, jasmine and forsythia. In nurseries of Ningpo he found a new yellow climbing rose, and in May he noticed clematis, spirea and azaleas growing wild over the hills. But Fortune soon concluded that botanists had already ransacked the south and looked forward with mounting excitement to the new plants he would kind as he travelled further north. Disappointments were in store for him, however. Near Tianjin the proprietors of several extensive nurseries welcomed him warmly. He was shown pots of dwarfed plants, including pears, Siberian crab apples and roses, all flowering profusely. Such techniques had only recently been attempted in the West, though they had been known for centuries in China. Other curiosities also came his way, such as several different species or varieties grafted on to a single stem. He remarked on the Chinese taste for trees that looked strange or old. But out of all he saw, the only cold-climate species were jasmines, weigela, roses and honeysuckles. All were already familiar to him, and all already introduced to the West by way of south China. The Chinese, he came to the conclusion, simply despised all northern plants as beneath their notice.

Changing tactics then, Fortune went up to visit the monasteries of the western hills beyond Peking. Here he was lulled by the soft tinkling of pagoda bells; he looked out over the dusty plain below and admired the huge pines, cypresses and junipers that grew in graveyards. The monasteries protected trees that were almost inevitably cut down anywhere else. Once he saw a fully grown *Pinus bungeana,* between eighty and a hundred feet tall, with peeling milky-white bark and fine light-coloured leaves. He realized only later that he had already sent home to England a seedling of the same tree collected in Shanghai. Even the monasteries could offer him nothing new.

In 1879 the next important collector to follow Fortune, Charles Maries, went westward up the Yangtze to Yichang in search of plants. There in the nearby hills he found *Primula obconica, Loropetalum chinensis* and the witch-hazel *(Hamamelis mollis),* and sent specimens of all these back to Europe. But he did not climb up far beyond the city; and like Fortune he concluded as he set off home that the floral wealth of China had already been fully catalogued by botanical explorers from the West.

However, way beyond the city walls and tended fields, much further than Fortune and Maries had ever penetrated, myriad forms of flora had in fact been growing for hundreds of years in unattended splendour, in places where there were 'nothing but mountains on every side . . . heaped one beyond another in quick succession and . . . separated by narrow defiles down which torrents rush and roar'. These are the words of Ernest Wilson,[1] who by exploring hitherto uncharted terrain, discovered a botanical paradise that had eluded his predecessors. He was one of the greatest of all the plant collectors ever to visit China. Like Maries he journeyed up the Yangtze to Yichang, but this was only his starting point. From there he climbed up into the mountains and looked back over a cultivated land of which every inch was terraced and worked. Ahead of him stretched a terrain so wild that even Chinese patience and ingenuity had not yet tried to tame it. As yet, however, the landscape was not entirely deserted. The tops of the cliffs were worn into grotesque and curious shapes, but all commanding peaks were crowned by Taoist temples, seemingly inaccessible, yet sheltered wherever possible by planted wintergreen, gleditsia, cypress, ginkgo and pine. As he moved further into these remote areas he stayed in solitary inns which commanded unbelievable views of cliffs and chasms. Around such places hardy rubber trees were cultivated and, at one particular house, so many magnolias that the air was heavy with their fragrance. Pressing forward, Wilson soon found himself surrounded by plants that had long since been banished from the cultivated fields below – plants he had not only never seen before, but which left him almost speechless. It was, he said 'simply a botanical paradise . . . beyond question, the richest temperate flora in the world'.

China owes the astonishing wealth of its plant life to a unique combination of geographical accidents. Even the mountain areas escaped the ravages of the great ice caps during the geological past, so that many plant species continued to develop in China that were wiped out in much of Europe and North America. Moreover, the foothills of the Himalayas, moistened by soft winds from the south, were an ideal environment for many alpine plants. Thirdly, in this warm and temperate climate three different floras – that of the colder, drier north, that of the sub-tropical south, and the alpine species – all mingled and fused freely for some thousands of

Karst peaks, Yangshuo, near Guilin, Guangxi.

Irises, Huanan (South China) Botanical Garden, Canton

years. One day, climbing high among precipitous limestone cliffs, Wilson collected thirty specimens of different woody plants. But more than anything it was the rampant growth and richness of it all that amazed him: the *Daphne genkwa* flowering 'everywhere on . . . bare exposed hills'; the Chinese wisteria looped about high trees; and the loropetalum spread so thickly over boulders that from a distance it looked like snow. Conspicuous on the hills were the bright yellow blossoms of *Caesalpinia sepiaria, Symplocos paniculata* with its abundant white fragrant flowers (later followed by brilliant ultramarine berries), *Anemone japonica*, irises, saxifrage, Henryii lilies, a lilac species which was new to him (*Syringa julianae*), a new hydrangea, a new genus shrub with yellow flowers (*sargentodoxa cuneata*), and a blue primrose which, 'descending in most shady places, . . . carpets the ground for miles'. Above all he was struck by the Banksia roses wreathing all the tall trees with their profusion of blooms.

From this earthly paradise Ernest Wilson sent the seeds of more than 1,500 different plants to England and the United States between 1899 and 1911. Altogether his collection numbered more than 65,000 specimens, representing about 5,000 species. Through his work alone and that of the nurseries for whom he collected, more than 1,000 plants were established in Western cultivation – all of them from the wild

Surprisingly, Chinese gardens seem to have remained totally unaffected by these discoveries. From then on collectors from England and America and other Western nations came and went, and shiploads of new finds were sent to Kew, St. Petersburg, Paris and Harvard. However, Chinese horticulturalists continued to use their skills to develop only those plants that had long since been domesticated; and in their gardens they mostly continued to love only those flowers that their ancestors had loved before them. Why had the Chinese not bothered to find all these plants themselves? And why, once they were found by others, did they not find a place for them in their gardens? The answer seems to lie partly in the conservatism that had gripped China since the beginning of the Ming era in the fourteenth century, and partly in the traditional Chinese way of valuing plants as symbols of ideas, moral qualities and emotional states, according to a continually developing body of associations that was first recorded in the *Shi Jing*. Plants discovered in the nineteenth and twentieth centuries, lacking this ancestry of symbolic allusiveness, tended to be thought of as unsuitable for the garden.

Flower Symbolism

Later teaching emphasized that it was the forms of nature themselves that had suggested ethical ideas to the ancient philosophers: trees and flowers were not merely illustrative of man's feelings and aspirations, but in some sense the source of them. Thus, in the eighteenth century the Qianlong Emperor wrote:

When I find pleasure in orchids, I love uprightness; when I see pines and bamboos I think of virtue; when I stand beside limpid brooks I value honesty; when I see weeds I despise dishonesty. That is what is meant by the proverb, 'the Ancients get their ideas from objects'.[2]

Often the ideas and the objects seem to the Westerner to be somewhat circuitously related. The orchids referred to here, for instance, were those tiny, modest, often greenish-coloured cymbidiums that came primarily from the south and west Chinese provinces. Like others before him, the Qianlong Emperor would have prized them for their graceful habit, for they were curved and balanced like the brushstrokes of a fine calligrapher. But especially he would have appreciated their scent. It was principally this that had suggested them from ancient times as symbols of uprightness. For the fragrance of orchids is discreet yet interesting, subtle and yet pervasive, so that it seemed to the Chinese like the friendship of an honourable man. Like the influence of such a man, this fragrance would steal across a room without anyone remarking it, yet when it disappeared each person present would feel its loss. In addition to this, orchids were difficult to grow, but if handled with proper delicacy and care might reward their cultivator with flowers uniquely blotched and stippled like a piece of fine Song porcelain. All these qualities endeared them to gentlemen and scholars, and the nurture and appreciation of orchids was potentially a moral as well as an aesthetic experience.

The Qianlong Emperor claimed that pines and bamboos made him think of virtue. Once again he was following conventional meanings, but the conventions were still rich enough in different associations to be a profitable subject to ponder as one listened to the wind rustling through their leaves and needles. Aged pines, gaunt and bent with the struggle to survive, were splendidly analogous with virtue triumphant. Bamboos, bending in the wind but never breaking, were a Confucian symbol of the true gentleman. Both are evergreen, and together with the lovely blossoms of the Japanese apricot (*Prunus mume*), which blooms on withered old branches even in the snow, they made up the celebrated 'three friends of winter'.

It is strange that the Emperor did not include in his list one other important plant which was a source of virtuous inspiration to a gentleman: the lotus flower. The lotus was seen by Confucianists as a model for the 'superior man'. Indeed, it is impossible to think of a lotus without recalling the lines of a famous eleventh-century scholar, Zhou Dunyi, which were memorized by every schoolchild in pre-Liberation China:

> It emerges from muddy dirt but is not contaminated; it reposes modestly above the clear water; hollow inside and straight outside, its stems do not straggle or branch. Its subtle perfume pervades the air far and wide. Resting there with its radiant purity, the lotus is something to be appreciated from a distance, not profaned by intimate approach.[3]

In Chinese this flower, the *Nelumbo speciosum*, is called the *lian* or *he* flower. *Lian*, though written differently, sounds like the word for 'unite', and *he* like the word for 'harmony'. The lotus therefore acquires an echo of both meanings when its name is spoken, and it is often used as a decorative symbol for friendship, peace and happy union. Apart from these meanings the lotus was also a popular Taoist symbol, the emblem of He Xiangu, one of the eight Immortals. However, it was exploited above all by the Buddhists, with whom it had been associated long before Buddhism came to China. On the open petals of the lotus the Buddha sits in paintings and statues all across the Orient, often with Bodhisattvas all around him on smaller or half-opened flowers. For Buddhists the lotus was the symbol of the soul struggling up from the slime of the material world, through water (the emotions), to find final enlightenment in the free air above.

Edward Schafer, speaking particularly of the Tang dynasty, has shown how even then the symbolizing habit in China fed upon traditional layers of association. The name of an exotic plant might be novel and exciting enough to make a colourful splash in a line of poetry, but without

RIGHT *The 'Three friends of Winter', pine, bamboo and prunus. Dish of Zhenghua mark and period (1468–87). Bamboos were considered the perfect subject for scholars' painting by the poet Su Dongpo, who also said of his garden, 'Better food without pork than a dwelling without bamboos'.*

RIGHT, BELOW *Gentleman appreciating lotus in a pavilion. In the lotus season the loveliest part of a garden was the pavilion set over the pond where the scent of the flowers rose from the water.*

the enrichment of traditional and historical associations it could hardly win a lasting place in the hearts of the people:

The plum [prunus], promising spring and the renewal of vitality and hope, or the peach, embodying fecundity and immortality in lore and legend, exemplify a host of old familiars, rich in human associations. Not so the lichee, for instance; though known to the north since Han times, even in Tang poetry it is still treated as an exotic, colourful and romantically charming, but only feebly expressive of ordinary dreams and passions.[4]

In the thousand years separating the new plant discoveries of the nineteenth century from the exotic imports of Tang this basic response developed into a much stiffer rejection of anything alien or new. Because of its unfamiliarity, the lichee lacked the power to deeply move even the receptive poets of the Tang. In the xenophobic late Qing, then, the discoveries of eccentric barbarian plant collectors were hardly likely to fare much better. And why should they? A flower without traditional levels of meaning was just an object, even if a well-formed, pleasingly coloured one.

In examining ancient sources such as the *Shi Jing,* it is hard for scholars to know if the ideograms for particular plants and trees refer to the species that they are used to describe today. Very likely the names have shifted their meanings, and the plants themselves have been so long under cultivation that they too have changed radically. Chrysanthemums are a classic case in point. In the *Shi Jing* the ideograph for yellow flower seems to have referred to chrysanthemums, but by the Tang dynasty (618–906) white blooms are mentioned, and even a rare purple kind. Under the Song (960–1279), chrysanthemum fanciers became so numerous that commercial growers vied with horticulturally minded scholars to develop more and more unusual varieties by crossing and grafting. In the twelfth century monographs began to appear on the subject: Liu Meng, in 1104, described thirty-five varieties; Shi Zhengzhi twenty-seven, all from the city of Suzhou; and Fan Chengda listed thirty-six, from the Suzhou suburbs. Many other kinds followed, and in the great flower encyclopaedia of 1708 no less than three hundred different chrysanthemums are included. They were given lovely names: 'heaven-full-of-stars' for a little branched, yellow-headed species that was similar to the wild *Chrysanthemum indicum,* and 'jade-saucer-gold-cup' for a large single white flower with a yellow centre.

Needless to say, these later types, bred for aesthetic reasons, were all light years away from the chrysanthemums known to the people of the *Shi Jing.* The plant had first been introduced from the wilds for its herbal qualities, and in ancient times it would have been eccentric to cultivate chrysanthemums purely for their yellow blooms. Use on the whole overshadowed beauty, and it was this emphasis that first brought trees into the garden.

Useful Trees

The people of the *Shi Jing* planted trees round their houses, as we learn from a poem in which a young girl begs her lover not to make a night-

time visit to see her:

> I beg of you, Zhongzi,
>
> Do not climb into our homestead.
>
> Do not break the willows we have planted.
>
> Do not climb over our wall,
>
> Do not break the mulberries we have planted . . .
>
> Do not climb into our garden,
>
> Do not break the hardwoods we have planted.[5]

'A Gathering of Literary Gentlemen in a Garden'. Detail of scroll attributed to You Qiu (fl. c. 1570–90). A scholar plays music under a banana palm, a symbol of scholarly ambitions.

Willow trees were probably used in those days, as they are now, for making ropes and baskets. Their leaves could be infused in water to make an inexpensive tea. And the leaves and bark of certain species were ingredients in preparations to relieve rheumatic pains, goitre, dysentery and bruises. In the west the healing powers of willows led to the manufacture of aspirin, a chemically produced version of the pain killer they contain. In antiquity the willow may well have afforded supernatural as well as physical shelter: in the Tang dynasty we know that the rebel Huang Chao, who sacked Chang'an in AD 880, spared only those families who hung certain leaves, including willows, over their doors. Buddhists later considered water sprinkled with a willow twig to be purified and holy. Later, various poetic levels of significance evolved, in addition to the practical and supernatural ones. The supple grace of willow trees, swaying in the wind like the waists of dancing girls, easily suggested them as similes for lovely women; the association of willows with water, and water with women, perhaps subconsciously reinforced this idea. Later, they also suggested sexual freedom and they were often symbols for singing girls and prostitutes. For this reason men did not plant them at the backs of houses, where the apartments of their women were. However, eighteenth-century Europeans who thought of them as typical of the Chinese taste were quite right to do so, for in gardens, on the banks of lakes and water courses, willows have retained an essential place.

The second tree mentioned in the *Shi Jing* – the mulberry – whose leaves are used to feed silk-worms – has never been widely planted in pleasure gardens. However, several of the most decorative trees still found in gardens today are anciently recorded as useful hardwoods. The particular type of hardwood Zhongzi was begged not to injure may well have been *Cedrela*

sinensis, the red cedar, which is now much planted as a courtyard tree because of its compact habit and edible young shoots. It was revered for its longevity, and in central and western China a man carried a mourning staff of red cedar on the loss of his father, just as he carried one of mulberry when his mother died. Alternatively, the hardwoods of the *Shi Jing* may have been catalpa trees, long prized for making coffins, chessmen, printing blocks and musical instruments, and for building houses. *Catalpa bungei* is especially grown as a shade tree in the north, and is much prized for its fragrant clusters of creamy bell-like flowers, finely marked with orange bands and purply dots.

Two other useful and highly ornamental trees which are recorded in ancient texts and still commonly appear in gardens are the paulownia (with a history going back some three thousand years) and *Sterculia platanifolia (Firmiana simplex),* known in China as *wutong,* the legendary perch of the Chinese phoenix. The paulownia, used as a source of wood for musical instruments, has large foxglove-like flowers in a delicate grey-blue colour, stained yellow at the throat. The *wutong is* celebrated for its edible seeds (eaten in moon cakes at the autumn festival) and for its fine large maple-like leaves of which the eighteenth-century poet Yuan Mei wrote:

> Half bright, half dim, are the stars.
> Three drops, two drops, falls the rain.
> Now the wutong tree knows of autumn's coming
> And leaf to leaf whispers the news.[6]

Wutong trees are constantly painted in Chinese landscape scrolls, nearly always sheltering a scholar's country study hall in company with an aged pine. In Peking in the Qing dynasty they were much grown in the Manchu palace gardens.

Inevitably, fruit trees were cultivated from early times; the persimmon, for instance, whose large round fruits of orange-gold (in China the colour of joy) turn soft and sweet with the first frost of Autumn, and the peach, prized as an ancient symbol of spring, marriage and immortality. A specimen in the gardens belonging to Xi Wang Mu, the Taoists' Mother Goddess of the West, was said to ripen every three thousand years, its fruits conferring immortality on those who tasted them. Pears could signify a more modest longevity, for the trees that bore them had been known to live three hundred years. The impartial Duke of Shao sat beneath a pear tree to administer justice in 1053 BC, so that forever after pear trees also suggested good government. On the whole these trees – and that favourite 'friend of winter', the prunus, were grown more for their blossoms than their fruit, while other favourite blossoming trees, like the much-planted cassia or osmanthus, were loved especially for their scent.

The classic ritual text entitled *The Rites of Zhou* (*Zhou li*) mentions five trees which were symbolically grown in cemeteries: pines for the ruler, thuja for princes, the summer-flowering koelreuteria for governors, *Sophora japonica* for scholars, and poplars for commoners. These applications did not in any way prevent these trees from being planted later in pleasure grounds. *Sophora japonica* – the 'pagoda tree' – is especially popular in gardens, and its pendular variety – the dragon-claw pagoda – with stiffly drooping branches that form a natural arbour, is widely cultivated in Peking. Pines especially were essential in a garden. Li Yu once compared a garden without an aged pine to a group of pretty women without a venerable man to set them off. Since neither pines nor thuja wither in winter they naturally stood for friends who remained faithful in adversity. It would be pleasant to attach this meaning to graveyard trees, but in graveyards both types were specifically planted to keep away the *wangxing* bird, which devours the brains of the unprotected dead.

During the Han and Tang dynasties, when Chinese culture was at its most receptive, new contacts were made with unknown worlds beyond Chinese civilization, and various exotic plants found their way into northern gardens. The pomegranate in particular, its red fruits opening to reveal innumerable seeds, became a symbol of fertility. In fact all through the Han, Tang and northern Song dynasties exotic trees and shrubs from distant provinces were sent up to the imperial capital to embellish the Emperor's parks, and some of these, though strange at first, eventually found a place for themselves in Chinese culture. From the south came tangerines, citrus fruits, and the lichee and loquat. It seems to have been in Han times too that the banana plantain first arrived in northern gardens from the sub-tropical south. Though at first it must have been merely curious, it soon acquired pleasing associations through a penniless scholar who was said to have used the wide leaves to write on for want of anything better. Thus the banana became known as the tree of self-improvement. It is now planted in gardens all over China, wherever it will survive in winter, carefully clothed with neatly fitted coats of thatch. Often it will be found by the garden study, for in particular it is valued for the pleasing, slightly melancholy sound of rain pattering on its broad leaves. In the Tang dynasty, as the tropical provinces of the far south became increasingly familiar to administrators of the Empire, other exotics began to be planted in the north, in particular the southern *gui*, which Edward Schafer identifies as the fragrant-flowered osmanthus.[7]

Flowers and Herbs

Magic was an important element in the ancient Chinese attitude towards plants and trees, and it became associated with particular kinds of flora either because they were medicinally useful or because they were uncommon. Holiness and medicinal value became inextricably linked, and this combination of powers perhaps also stirred the aesthetic faculties of herb gatherers: a medicinal leaf, which was therefore a holy leaf, began also to be thought of as beautiful. Exceptional blossoms, whether on shrubs, trees or herbaceous plants, were everywhere interpreted as omens. An unusually vigorous shrub, or an orchid which flowered more profusely than expected, would be a direct sign of propitious forces at work for the benefit of the owner. The same principle also operated in reverse. In the Tang dynasty, when Emperor Ming Huang fled to Shu, the southern tangerines no longer flowered in the imperial park. In the *Dream of the*

A scene in the Yuan Ming Yuan with a wisteria trellis, painting by Tang Dai and Shen Yuan, eighteenth century.

Red Chamber, the family's impending collapse is presaged in the garden by a begonia which suddenly bursts into flower in mid-November. All the important members of the household go out to inspect this phenomenon, and although Baoyu's strictly Confucian father sees it merely as an interesting botanical event, everyone else is very much disturbed. Reversals of nature cannot but bode ill, for they reflect an inner lack of harmony which sooner or later will lead to disaster.

While unusual botanical phenomena foretold extremes of good or bad fortune, modest well-being was signified by the vigorous but unexceptional blossoming of common plants such as the chrysanthemum – rich in magic juices and beautiful in form. Chrysanthemums were first grown for their medicinal properties, and became an important ingredient in the life-prolonging elixir of the Taoists. In ancient times, one story went, people of Nanyang in central China drew their drinking water from a stream where chrysanthemums grew. Essences from these plants seeped into the water, and the Nanyang residents all lived to be a hundred. The unusual flowering season of chrysanthemums perhaps promoted their connection with long life, for they bloom in the autumn when everything else is dying off. Tonic wine was brewed from an infusion of their petals, and fragrant chrysanthemum tea was good for the health.[8]

Chrysanthemums were only one among many types of ornamental flowers that first came into the garden by way of a pestle and mortar or the cooking pot. In the Tang dynasty several poets wrote of the lotus-gathering ladies of the south, whose graceful figures, leaning out among the scented blooms, could still at that time add an exotic flavour to a line of verse.[9] They were not, however, ladies of leisure, but hard at work harvesting a crop. Lotus are among the most useful of all decorative plants. Their tubers, eaten raw or cooked, are sweet-tasting, crisp and juicy. A starch can be made from them which is readily digestible and thus often given to the sick. The seeds are added to soups, or used to make a sweet paste, often used in mooncakes. The leaves are used for flavouring, and to wrap things in. Yun, the delicious wife of Shen Fu, in *Six Chapters of a Floating Life* (the eighteenth-century memoirs of an impoverished scholar), used to make little gauze tea-leaf bags, which she tucked into the open hearts of lotus blossoms. The petals closed in the evening, trapping the teabags inside, and by the morning they would be delicately scented with a lingering fragrance of lotus.[10]

There is one flower, however, which arrived in gardens relatively late and rapidly became a favourite: the *mudan,* or tree peony. It is the embodiment of aristocracy, wealth and rank, and beautiful women, and also, paradoxically, of the principle of *yang* (*mudan* literally means 'male vermilion'). Interestingly, this plant is not mentioned in the *Shi Jing,* or the *Book of Rites,* or the *Li Sao* (the main poem in the *Chu Ci*). It is not mentioned in the literature of the Han dynasty. Its earliest reference seems to have been in the fourth century, in the writings of Xie Lingyun, China's first landscape poet. He was fond of mountain hikes, and the tree peony he describes, with its glorious profusion of blooms, seems to have been growing in the wild. No mention is made of such flowers in a garden until the Tang dynasty when they are described by the poet Li Bai spectacularly opening out their reddish-purple, pink and white blooms for the famous beauty Yang Guifei in the park of the Emperor Ming Huang.[11] Even their luxuriant and finely-shaped leaves were in themselves pleasingly ornamental. In later times the rough bark of the tree peony would be prescribed for blood disorders, but its first appearance in the imperial gardens suggests

LEFT *Peonies under an awning in the Zhi Garden, by Zhang Hong (1627). The peonies are displayed in a special bed where they can be admired from a fine pavilion.*

RIGHT *Peach blossom in the Imperial City, Peking.*

that it was one of the earliest flowers brought into cultivation for pure show.

Because it had so much to offer, even the Emperor could not keep exclusive rights to the tree peony for long. It began to be grown in great private mansions, and soon was flourishing all across the city of Chang'an. A peony-growing industry developed, producing ever finer and more unusual variations, but even the most ordinary bloom stood for wealth, riches, and honour. Under Empress Wu Zetian, who reigned from AD 684 to 705, the centre for the cultivation of peonies moved to Luoyang, where double-petaled varieties were introduced. During the flowering season a Festival of Ten Thousand Flowers was annually declared. The four finest blooms were chosen from all the exhibits, wrapped in cabbage leaves with their stems sealed in wax and packed in bamboo cages to be sent off to the Empress by relays of horsemen. The court waited to see them, breathless with anticipation. Meanwhile, back in Luoyang, the whole population crowded round to admire the rest of the displays, although few could afford to buy the finest. In the early ninth century Bai Juyi described a similar flower festival as it was in his day: awnings were set out to shade the blooms and there were tantalizing glimpses of special varieties, half displayed to the crowds and half hidden behind wattle fences.

Cheap and dear – no uniform prices,
The cost of the plant depends on the number of blossoms
The flaming reds, a hundred on one stalk;

The humble white with only five flowers . . .
If you sprinkle water and cover the roots with mud
When they are transported they will not lose their beauty.[12]

Even the poor went along but it was principally an urban pastime. When an old farm labourer in the poem hears the price of a 'cluster of deep-red flowers', he sighs, because the amount would have paid the taxes on 'ten poor houses'.

The love of peonies rose to be a kind of national passion in the Song dynasty. Ten thousand differently coloured bushes were ordered for the imperial summer palace at Li Shan. The man in charge was Song Shangfu, a poet as well as peony-cultivator, for by this time the ability to grow fine varieties of *mudan* was a valuable accomplishment for any gentleman of taste. In time the fame of Luoyang as a source of beautiful peonies was displaced by the rise of other important centres of cultivation. In the Southern Song, Lu You wrote a monograph on the Tianpeng *mudan* of Sichuan; while under the Qing, Bazhou in Anhui became famous, then finally Caozhou in Shandong and some of the cities near Shanghai and Lake Tai.

Plants in the Garden

The essential role played in Chinese culture by plants, especially in Song times, is obvious from the vast literature either wholly or partly devoted to them. Lists of plants were compiled giving all their medicinal uses, while 'famine herbals' described the edible parts of every known tree and flower. Horticultural monographs were produced on topics such as the cultivation of orchids, and extravagant eulogies claimed to prove once and for all the superiority of the lotus, or the peony, or the chrysanthemum, over all other species. Meanwhile, through poetry and prose the symbolic and expressive meanings of flowers were continually enriched and expanded. However, while there is no doubt about the importance attached to plants in China, they do not at first seem to be given any special prominence in the making of gardens. This is startling to anyone who is accustomed to the wealth of plants in a modern English garden. An old Chinese jingle lists the attributes of a perfect garden like this:

> Breezes in spring,
> Flowers in summer,
> Moon in autumn,
> Snow in winter.

Out of the four seasons, only summer is thought to be important for plants and then only for flowers, with no mention of trees and shrubs.

Nevertheless, growing things are in fact deployed in Chinese gardens in subtle and important ways. By chance, my own first visits to the gardens of Suzhou were in March and early April, when more than half the trees were still bare. Through the fine sunlit network of trunks and branches, the juxtapositions of walls and rocks, water and pavilions, seemed infinitely more fascinating than the rich planting of an English garden. Evergreen pines, groves of rustling green bamboos casting shadows on the white walls, and pale budding fronds of willows blowing in the breeze like strands of yellow hair, were all thrown into sharp focus by their isolation from other signs of growth.

That same April I was amazed to see a magnificent rock in the back garden of the Imperial City in Peking; though it looked from a distance like the craggy peak of some inaccessible mountain range, it was still entirely covered with the dried grey arms of last year's creeper. In winter it might be a symbol of uncompromising wilderness, but in summer it would almost disappear beneath a thickly undulating blanket of foliage, and presumably the same thing would happen to the rest of the garden as the year wore on. All the intricacies, the delicate balance of different elements, the dappling of reflected light on walls and rocks, would be lost in a more ordinary sensation of trees and flowers.

It was only much later, when I eventually saw Suzhou gardens hazy with fruit blossoms, that I began to appreciate the mysterious expansion of space accomplished by the planting of a Chinese garden. As the trees unfold in all their different greens, each courtyard and space cell in the garden becomes more intensely private and enclosed, yet the effect is anything but claustrophobic. As the leaves and branches move in the breeze, and the shadows below them shimmer, it is not always easy to tell if the white walls are real or illusory. Through the open-work windows of galleries and walls, bright splashes of shrubs in flower highlight parts of the garden that passed unnoticed before. And the scents borne on the wind make streams and pockets of fragrance that complement and counterpoint the flow of walls and rockeries, of open vistas and enclosed courtyards.

More than any others, private gardens in China celebrate change. It is illuminating to make a comparison with English landscape parks. In winter and summer an English park such as Stourhead remains basically the same: the colours are different, and the density of the forms, but not the structure of the various views. At Stourhead the foliage and flowers are part of the garden in much the same way as a chameleon's skin is part of its body. In a Chinese garden, however, the growing trees and flowers are more like clothing, sometimes enhancing, sometimes obscuring, and sometimes totally altering the effect of the forms beneath.

In particular, seasonal change is intensified in Chinese gardens by the way in which different courtyards and space cells are often used in rotation to celebrate the flowers of different seasons. Peonies, for instance, are usually planted all together in beds edged with stone or marble, or in rocky terraces raised on the side of an embankment. In early summer, when the flowers are in full bloom, that part of the garden becomes a main centre of attention. The peonies are not cut, but grow, bloom, fade and die all in their own place. Visitors come to do them homage. In imperial China, parties were given in their honour. Host and guests sipped wine, improvised verses and joked together as they sat looking out over the showy blooms before them. However, when the peony season had passed, all social activity in the garden moved on to another focus.

As the summer heat increased, most of the garden visitors would spend their time in open pavilions set on stilts over corners of the lakes planted with lotus. These airy *ting*, shaded with gauze curtains, would catch the first evening breeze. By leaning out over the curved balustrades, visitors could just touch the leaves of the lotus plants growing up from the water below. A lady of the sixteenth century, Huang Youzao, describes the languorous *ennui* of such summer afternoons:

> In the deep hall the dust dissipates, the noon tide heat fades;
> Smoke curls like a dream, the day is lagging, lagging –
> A light wind seems in treaty with the lotus:
> Bearing the fragrance along, it rolls up the curtain.[13]

RIGHT   *The lotus lake in early spring, Zhuo Zheng Yuan, Suzhou.*

RIGHT, BELOW   *The lotus lake covered by its summer growth of lotus leaves, Zhuo Zheng Yuan, Suzhou. Lotus effect an extraordinary change in the garden between spring and summer when they often completely cover the water of a pond with their luxuriant leaves. In winter the bent and broken stems remain, reflected in the water on still days like characters written in fine calligraphy.*

Lotus – the flower of summer lassitude, just as the peony is the flower of early summer plenty – causes one of the most extraordinary seasonal transformations in the Chinese garden. In spring the lakes are glassy reflecting pools, but in the lotus season the leaves of these incredible plants rise high on their curved stems to make a new billowing, green surface several feet above the water. Lorraine Kuck describes them blooming in Peking's Beihai Park during the 1930s:

> The leaves are shaped like an immense rounded bowl with rippled edges. The interior of the leaf is
> a curious translucent, bluish green, due to a waxen surface. When drops of rain or dew collect in
> these jade bowls, surface tension pulls up the water into crystalline globules. And if the leaf sways,
> these jewel-like drops roll back and forth like crystal beads. When acres of those leaves rock together
> in a light breeze, sheets of electric blue light sometimes flash across the field, refracted from the
> waxen surface.[14]

In the larger lakes, like the one at Beihai, paths used to be cut through the lotus beds so little punts could pass through them, invisible beneath the vegetable surface. Houses without gardens would grow lotus in their courtyards in large bowls filled with fine, rich mud and topped up with water. Such bowls were often placed on either side of important entrances, and the huge bluish leaves would fan out above them like little sunshades.

The boundaries of China enclose a vast range of climates. In the north, peonies sometimes replace lotus as the traditional flower of summer, while in the south they are the flower of spring. North and south alike, however, in temperate or subtropical zones, the symbol of autumn is everywhere the chrysanthemum, and inevitably when chrysanthemums bloomed it was a fine

The petals of a lotus, the flower of high summer, glow in the light. As the summer heat increased, garden visitors would spend their time in open pavilions set on stilts over corners of the lake planted with lotus.

Pot garden, visitors' rest house, Guangdong province.

excuse for seasonal entertainments, however humble a person's circumstances. Shen Fu and his wife Yun, the destitute couple of *Six Chapters of a Floating Life*, spent another hot summer at a tiny cottage surrounded by fields in the suburbs of Suzhou. As usual, sensibility triumphs over poverty in the most enjoyable way:

> I had asked our neighbour to buy us some chrysanthemums and plant them along the bamboo hedge. When the plants began to bloom, in the ninth month, [we] decided to stay for another ten days or so to enjoy their beauty and to send for my mother to come and see them also. The invitation pleased my mother and after she arrived we spent the whole day in front of the flowers, eating crab-legs and gossiping.[15]

As well as being planted in the garden itself, chrysanthemums, together with other seasonal flowers, would also be grown in pots. In every large, elegantly designed garden herbaceous plants and dwarf trees would also be grown in pots, in a special area set apart behind a wall or wattle fence. When in bloom the best plants were taken out to be displayed on tables and flower-stands in each of the garden's halls and studios, placed in courtyards, among rocks, on open terraces and on either side of entrances. Corridors and low walls throughout the garden would be decorated with such pots brought out in rotation. In this way a foreground of constantly changing forms and colours was provided which supplemented the more permanent trees and shrubs. As one walked along the winding open galleries the flowers were close enough to be inspected individually, but they also combined with the architecture to frame the view of the garden beyond. Moreover, the designs of wooden lambrequins and balustrades echoed the shapes of the flowers and leaves placed near them, so that the flowering plants in their pots became a subtle but important link between the natural and manmade elements of a garden.

In the private gardens of the rich these decorative pot plants were naturally cultivated by professional family gardeners. However, certain flowers were also often grown personally by scholarly horticulturalists. Chrysanthemums were among them. Thus, as well as public chrysanthemum shows, there were also in imperial China private gatherings of gentlemen fanciers who each brought a pot or two of their most successful blooms to compare with those of their friends.

An artistically inclined gentleman such as Shen Fu would have been well trained in the many nuances of flower appreciation. Judging flowers of any kind was always a delicate and lengthy business. In *Queen Mary and Others,* Osbert Sitwell describes a scene that took place in Peking in 1934, after the end of the last dynasty, when the old hereditary Manchu aristocracy came out for the first time in many years to view the crab apples blooming in the garden of one of the few surviving imperial princes. The old men, as bent and contorted with age as the two-hundred-year-old trees they had come to inspect, climbed painfully up stone steps to look down on to the blossoms from above.

TOP *Gardenia, by Xu Xi (tenth century).*

ABOVE *Basket of Flowers, by Li Song (1166–1243).*

RIGHT *Thatched pavilion in spring in the park outside the city walls, Nanking.*

> Once there they would remain a full hour, matching in their minds the complexion and fragrance of the blossom of previous years with that before them.... Critical appreciation of this high order could not be hurried. After all, it was better fully to use now the powers of judgement with which the years had enriched them, and to apply their trained abilities in this direction, for in the order of things, they could scarcely hope to see many more of these flowery harvests.

As Sitwell left he looked back and saw each old man outlined against the blue Peking sky, gazing down into the 'frothy intricacies below, and waiting, quietly, like a watcher on a tower or the guardian of an ancient shrine, calling the faithful to worship'.[16]

As well as chrysanthemums, literary horticulturalists also liked to cultivate the little cymbidium orchids (known as *lan hua* in China) which, as we have already seen, made the Qianlong Emperor think of uprightness. These too were grown in pots. There were two main varieties; and the one that bloomed in spring was everywhere regarded as the symbol and harbinger of the season.

Feelings ran high over the cultivation of orchids and gentleman growers sometimes cracked under the strain. Shen Fu, for instance, inherited a particularly beautiful and rare plant on the death of an old orchid-fancier friend. It was a spring orchid 'of the lotus type, with broad white centres, perfectly even "shoulders", and very slender stems'. After he took it home, Shen Fu 'treasured its perfection like a piece of ancient jade'.[17] It flourished under his care, but one day suddenly died. Sad and bewildered, Shen Fu dug up the roots to see what had happened, and found they had turned quite soft. Later he discovered that a rival, to whom he had refused a cutting, had deliberately poured a pan of boiling water into the pot.

The writer and painter Su Hua (Ling Shuhua), remembering her childhood at the turn of the century, tells us that there were more than twenty kinds of orchid in her family's greenhouse in Peking, all of them coming originally from the south; of these the most valuable was the *Chien*

lan (*Jian lan* or Fujian orchid), with its light-green flowers and hearts. One very rare variety owned by her family was the *Ouk-lan* (*Yue lan* or Guangdong orchid), which had been brought to Peking from Guangdong. The Shandong orchids from the eastern peninsula, with their greeny petals and red hearts, were common in all the greenhouses of the capital. And many plant lovers also possessed the lavender-petaled kind of hanging orchid that had been imported from Sichuan in the far west.

Every morning after her lessons, Su Hua went to water the orchids with the family gardener Lao Chou (Zhou). Her descriptions of him, as a wise and loving man, are also an intimate portrait of gardening life in early twentieth-century Peking. As a treat, the little girl's mother would allow her to go with Lao Chou to the city flower market. There they would collect parings of horses' hoofs from friendly smiths, to make into a broth for feeding the orchids. They also bought offal which Lao Chou buried later near the roses in their garden. Next, they would visit some of the gardener's old apprentices, who now ran commercial greenhouses in the market. Lao Chou had helped the young men by giving them cuttings to start their greenhouses, and they in turn would now do their best to find unusual plants for him.

On one memorable holiday, Lao Chou took Su Hua to visit a white-haired old countryman who had been head gardener at the Summer Palace in the days of the old Empress Cixi – whose touching concern for the welfare of her plants had been balanced by a remarkable indifference towards human beings. His memories give a good idea of how flowers were used in the Summer Palace:

> Every day I had to remember to send some fresh new flowers to the pavilions where the Empress might be. Only God knew which pavilions she was going to visit. To play safe I had to send to most of them. If the Empress disliked the flowers she would send for me. She wanted me to grow all the flowers and blossoms that figure on pictures and scrolls. She had very unusual taste in flowers. She could never bear the curled or trained flowers that we see in the market. So I had to grow all the flowers in our gardens and our own greenhouses.[18]

In the old man's garden Su Hua saw a two-hundred-year-old wisteria, 'its trunk like a dragon rising from the sea'. She was also shown a lilac which was three hundred years old and reminded her of 'a woman dancing with a beautiful motion'. Such gems he tended with infinite love and care, following a body of gardener's lore which had been passed down orally from generation to generation. He drew from this rich and ancient tradition not only shrewd practical details of horticulture – for instance, knowledge about what oils to apply to rotting roots, and how to propagate fruit trees – but also a special feeling, inspired by the Taoist beliefs, about a gardener's relationship with his plants.

In Chinese gardens it often seems as if the side effects of flowers and trees – and indeed of architecture and even rocks and water – are almost more important that the objects themselves. The delights of a garden lie in the freshness that rises from a stream in the summer heat; in the distant sound of a fish leaping in the evening mist; in the scent of lotus and the transforming power of snow. But in particular they lie in that peculiarly Chinese fascination for paradox – for moments when the buzz of a cicada (as one saying goes) heightens the silence, or a rustling breeze make the stillness more intense. By immersing himself in these transient but often-repeated effects, the garden maker grew in awareness of the Dao's eternal transformations and, in the acute perception of the passing moment, himself transcended time.

MEANINGS OF THE CHINESE GARDEN

by Charles Jencks

LEFT *A quiet corner in Suzhou.*

ABOVE *Lou Lim-Ieoc Garden, Macao. Portuguese and Chinese taste mixed in a merchant's subtropical garden.*

It would be vain, and probably wrong-headed, to try to summarize all the different meanings of Chinese gardens which we have touched on. They do not add up to any single conclusion, and there is no one type that is the essential one. The tradition was never intended to be summarized and turned into design formulae. Indeed, Chinese gardens are built today which are as various in their approach as to seem opposites. In Cheshire, Connecticut, there is one in the tradition of Wang Wei's country retreat built by Nelson Wu, a scholar whose fascinating views we shall be discussing shortly; there is a delightfully vulgar recreation of an Immortals' Isle in Taiwan; and on the mainland several of the most famous old gardens have been restored while many new parks have also been laid out.[1] None of these is very like another, and yet they can all claim to be extensions of the Chinese tradition.

Nature Versus Artifice

Yet if there is no single, grand conclusion to the history, nevertheless there are a series of recurrent themes we have seen reappearing like leitmotifs in each chapter, and these point to underlying ideas. One of these ideas is that of naturalism. How natural are the winding paths, the fantastic grottoes, the re-created farmhouses? This question, seemingly unanswerable, forms the subject for a major dispute in the *Dream of the Red Chamber,* a dispute which throws light on the difference between the natural and the artificial. Many of our themes come together around this distinction: the idea of the country farmhouse which symbolizes Confucian simplicity, the idea of aesthetic convention, the idea of *qi* and the need to be in accord with the 'vital spirit' of nature, and the notion of the Dao.

The dispute takes place significantly enough in a newly built garden, when the head of the family Jia Zheng and his son Baoyu are making their first inspection along with a group of literary gentlemen. The discussion is set off by the discovery of a mock country village – or should we say realistic thatched hamlet? – in the midst of the vast and lavish pleasure ground. The hamlet, which sits amongst several hundred apricot trees, consists of a little group of reed-thatched cottages. Jia Zheng reacts to these as a proper Confucian: he is relieved to find such an unpretentious vision among all the frippery. 'Ah!' he exlaims, 'now here is a place with a purpose. It may have been made by human artifice, but the sight of it is none the less moving. In me it awakens the desire to get back to the land, to a life of rural simplicity. Let us go in and rest awhile!'

His son Baoyu, however, has a fresh eye on these things and quite a different notion of naturalness. He is young, lively-minded and not yet properly schooled in the accepted forms – or perhaps he is just sceptical of his training. First of all, he says, he really prefers the pretty painted pavilions they have already seen. His father is annoyed. 'Ignoramus! You have eyes for . . . only the rubbishy trappings of wealth. What can you know of the beauty that lies in quietness and natural simplicity? This is the consequence of your refusal to study properly.' Baoyu replies with the provocative statement that he has never really understood what it was the ancients *meant* by natural: his father, perhaps enraged by this baiting, makes the fatal oversimplification for which the son is hoping: '"Natural" is that which is *of nature*, that which is produced by nature as opposed to that which is produced by human artifice.'

Baoyu easily disposes of this view: 'There you are you see! A farm set down in the middle of a place like this is obviously the product of human artifice' – as indeed is everything in the garden, no matter how naturalistic it may look, or how much nature may alter it over time. But Jia Zheng has granted this much already ('It may have been made by human artifice . . .'), and so Baoyu must attack the deeper notions of conventionalized nature, not merely his father's momentary lapse. Thus he questions the plausibility of the village in this elaborate context, and the degree to which it simulates villages he knows: 'There are no neighbouring villages,' he says, 'no distant prospects of city walls; the mountain at the back doesn't belong to any system. . . . It sticks up out of nowhere, in total isolation from everything else.'

In short the hamlet is unnatural or artificial because it is incongruous, inconsistently carried through, especially when compared with the 'natural form and spirit of the places they have already seen'. What kind of distinction is Baoyu making? Two different translations of this key passage bring out the problems. David Hawkes' version continues this way:

> The bamboos in those other places may have been planted by human hand and the stream diverted out of their natural courses, but there was no *appearance* of artifice. That's why, when the ancients use the term 'natural', I have my doubts about what they really meant. For example, when they speak of a 'natural painting', I can't help wondering if they are not referring to precisely that forcible interference with the landscape to which I object: putting hills where they are not meant to be, and that sort of thing. However great the skill with which this is done, the results are never quite. . . .'[2]

Just as we are about to hear the concluding point, Jia Zheng shouts his son down, and the argument is cut short. However, Andrew H. Plaks glosses Baoyu's statement here as 'a dislike of forced statement, a deliberate play for naturalness, and not the spontaneous suchness to which the Chinese term for naturalness (*tianran*) so aptly refers'.[3] He goes on to cite the garden manual *Yuan Ye,* which warns against violating the principle of instinctive naturalness (*de ti*). However, as we have seen in the chapter on painting, there are two distinct but always mixed-up views on what constitutes the fundamental principle of *qi yun sheng dong.* It is a principle of inner creative spontaneity, the 'life-breath' within the painter, and at the same time concerns the 'resonance' of this with outer nature. It is a rule of both psychology and realism, of the creative individual and creative nature.

Baoyu is using this mixture of meanings not only against his father's conventional Confucian idea of simplicity, but against painters' conventions as well. He finds them all implausible when compared with his own understanding of nature. He might even have agreed with Oscar Wilde's urbane put-down, 'To be natural is such a difficult pose to keep up'. A similar battle of the conventions was taken up independently, with great enthusiasm, by garden theorists in England during the eighteenth century who helped form a taste for more irregular arrangements,

'Enjoying antiques' by Du Jin, fifteenth century. In Chinese households valuable antiques were not displayed permanently, as they are in the West, in cases or on shelves, but were brought out for their owner to admire and appreciate with his friends, then put away again. Here a successful man has had an unofficial portrait made of himself in his garden, where his collection of bronzes has been placed on a table. A screen protects his back from the breeze; overhead trees provide shade, and the garden is enhanced by the presence of ancient objects.

supposedly more natural to the English countryside. Yet the models were still 'artificial' – the paintings by Lorrain and Poussin of the Roman campagna! All gardens must be artificial arrangements, like most of the plants grown in them, if 'natural' is taken to mean 'in an original state'. The controversy is therefore a pseudo-war. Yet it is true that the English and Chinese share a penchant for 'consulting the genius of the place in all', and for a picturesque heightening of natural effects. Both used the naturalistic argument when it was convenient to do so, and conveniently disregarded it when discussing such obvious artificialities as fantastic rockeries and unearthly grotesques and grottoes. That such a contradiction should be so blatant and so long-term indicates that it cut across several deeply held values.

To be 'natural' was of course to follow the Dao of nature, getting oneself in tune with the underlying rhythms of the seasons, the plants, the universe, so that there was no discrepancy between inner being and outer reality. Since this was the ideal for painters too, as well as, in a different sense, of the Confucian *li* (the right principle of order, or conduct), then a lot was at stake when Baoyu and his father both appealed to their vision of the 'natural'. We are witnessing a fundamental philosophical dispute between on the one hand the Confucian ideal of simple rusticity, the Taoist ideal of natural spontaneity, the painter's ideal of 'vital spirit'; and on the other the several conventions for representing these ideals. It is not at all surprising that the dispute should take place in a garden, for after all a garden was precisely that microcosm of the

universe where all forces would be present, or at least represented – an idea which takes us to the next most important theme underlying a Chinese garden.

Plenitude and Restless Polarizing

As we have seen, the garden was used and thought about in many ways, some of which were contradictory. The profusion of contrasting uses seems at first to defeat any attempts at summary and understanding until we focus precisely on this superabundance and opposition. Then the kaleidoscopic patterns begin to become clear. The garden, as a microcosm of nature's plenitude, must pack all experience within it on an aesthetic level, and this packing creates a peculiar kind of space, which we will consider shortly.

A garden would be used for solitude as well as entertaining friends, for study as well as the occasional dalliance, for quiet intoxication as well as cultivation, for composing poetry and meditating as well as family outings or boat-parties and, on an Imperial scale, even for war games. Opposed to life in the Japanese or English garden, which is more contemplative and passive, life in the Chinese garden is a mixture of everyday use and contemplation. Since the Chinese garden is a symbol of the owner's life and character, it must express and articulate his day-to-day activity as well as his abstract thoughts. There is thus a function given to the garden which it cannot completely fulfil: it should symbolize the entire universe with its 'ten thousand things', and make this profusion quite intelligible. Andrew Plaks takes this to be a fundamental

LEFT *Consequa's garden in Canton, engraving from a painting by Thomas Allom (1804–72). A fantastic – and rather vulgar – garden belonging to one of the Chinese merchants who carried on foreign trade in the eighteenth century. The leaf-shaped roof on the pavilion is worth noticing, as are the live deer and water fowl. The artist has merged this highly stylized pleasance into naturalistic woods in the background.*

RIGHT *Enjoying life at home, nineteenth-century woodcut from the* Hong Xue Yin Yuan Tu Ji *by Lin Qing. The Inspector general looks up from his table to watch his little son chasing a butterfly with his fan. Around the open hall a small courtyard has been enclosed, so the scholar has a pleasant and soothing outlook while still enjoying seclusion.*

role of all literary gardens: 'The enclosed space of all literary gardens clearly "stands for" the sum total of finite creation.'[4] But this gives rise to a problem, for although the garden tries to represent the total intelligibility of the world, and the idea that everything is knowable, it also quite clearly rejects a literal-minded grasp of this knowledge.

The way in which the garden resolves this insoluble dilemma is twofold: it tries to incorporate a bit of every experience within a tight space; and it restlessly oscillates between polar opposites, the *yin* and *yang*, the solid and void, or what Plaks has shown to be another archetypal pattern, the five elements – meaning in general 'many'.[5] In short the garden symbolizes the universe through its formal devices. This partially explains what is so characteristically strange to us about the Chinese garden: its cramming of a density of meanings into a very small space, its tight packing, and its restless changing aspect. So ceaseless is the alternation in moods and vistas that it cannot be totally accounted for in aesthetic terms. A well-known passage from Shen Fu, the eighteenth-century writer, connects the aesthetic motivation with the symbolic.

> In laying out garden pavilions and towers, suites of rooms and covered walkways, piling up rocks into mountains, or planting flowers to form a desired shape, the aim is to see the small in the large, to see the large in the small, to see the real in the illusory and to see the illusory in the real.[6]

In other words, one thing is a substitute or symbol for another – the part of the garden is standing for a larger part of the universe. A gentleman well versed in such symbolism would even appreciate how absent qualities were represented by present ones, how the autumn moods implied the next spring, how every visible quality – rocks, voids, colour – implied its counterpart

– water, solids, monochrome. This symbolism of negation is a recurrent theme, as we have seen in the attitudes of emperors as well as painters or literary gentlemen: all have a well-developed taste for the autumnal, for the aged tree, for the insipidly empty, and part of the reason is that these negative qualities all symbolized their opposites. In any garden the polar opposites are designed so that one is always waxing while another wanes. If there is too much of one quality the designer must swing the pendulum back the other way, not so that it comes to a rest, but so that it goes on swinging. The continuation of Shen Fu's passage brings out this restless polarization:

> Sometimes you conceal, sometimes you reveal, sometimes you work on the surface, sometimes in depth. One need not waste labour in vain merely in terms of the well-known formula: 'winding, deceptive intricacy' or 'a broad area with many stones'.[7]

The job of garden building, like the builder's character, is never completed, but always undergoing growth, decay and transformation. And yet such symbolic justifications do not entirely account for the formal profusion and intricacy so characteristic of the garden: there are also other motives at work.

Magical Space

If the garden is a serious attempt to symbolize the universe, and indeed the Land of the Immortals or the Buddhist paradise of Amitabha which lie behind it, then its structure should share some aspects with other religious forms or ritualized spaces. As we have seen, the origins of gardens were bound up with holy mountains and magic rocks, some actual (and disastrous) attempts at immortality on the part of Emperors, and a long unbroken tradition of scholars and recluses trying to attain harmony with the mystery of the Dao. Some recent writers have claimed that the Chinese love of nature was so strong, as evinced in Chinese landscape painting and gardens, as to constitute a philosophy, or belief, on a par with Confucianism, Taoism and Buddhism.[8] Thus, it is fair to look at the garden plan as a literal religious form, and with this perspective in view to ask what type of experience it created.

Edmund Leach, in a recent study of various kinds of communication, has shown the typical structures of forms concerned with myth, ritual, taboo and religion.[9] He finds underlying these mytho-logic forms a pattern quite different from the everyday logic of conventional events, a pattern owed to the interruption of normality, the transcendence of boundaries. In his terms, the Confucian house with its sharp boundaries made by high walls, its ordered symmetry and social hierarchy carefully articulated in plan, would be juxtaposed with the Taoist garden that has deliberately confusing inner boundaries, no clear order, and certainly no symmetry.

Leach concentrates on the crucial notion of the way in which man introduces artificial boundaries into all his activities, arbitrary divisions such as those which cut the continuous flow of time into seconds, hours and days. These divisions order the natural, or biological flow of life into discrete units, which now acquire social and religious significance. One can thus speak of social time interrupted by the rituals which mark the transition between each discrete social state: for instance puberty rites, weddings, funerals and healing rituals. The boundaries which mark each clear-cut social state, say the interval between unmarried and married, are fraught with anxiety and experienced as timeless, ambiguous and sacred. We might term a Chinese garden 'spaceless', for the analogous reason that, with its abnormal and incomprehensible patterns, it interrupts the normal social and functional relationships of the city.

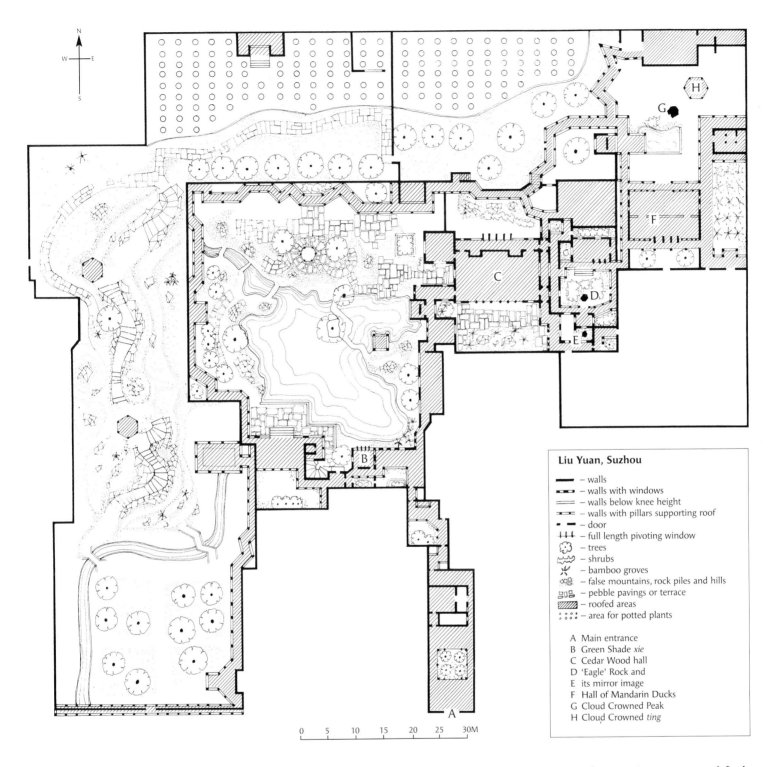

Liu Yuan, Suzhou

- ▬ – walls
- ▬·▬ – walls with windows
- ═══ – walls below knee height
- ▬·▬ – walls with pillars supporting roof
- ▬ ▬ ▬ – door
- ┼┼┼ – full length pivoting window
- – trees
- – shrubs
- – bamboo groves
- – false mountains, rock piles and hills
- – pebble pavings or terrace
- – roofed areas
- – area for potted plants

  A Main entrance
  B Green Shade *xie*
  C Cedar Wood hall
  D 'Eagle' Rock and
  E its mirror image
  F Hall of Mandarin Ducks
  G Cloud Crowned Peak
  H Cloud Crowned *ting*

0   5   10   15   20   25   30M

Leach goes on to describe the role of the mediator between these various states, and finds an archetypal structure, whether this mediator is a prophet, a ritualized form of behaviour such as sacrifice, or an actual building such as a church. The common aspect of the mediator in all these cases is that he, she or it takes on 'liminal attributes – being both mortal and immortal, human and animal, tame and wild'. The Biblical prophets who mediate between the City of God and the City of Man occupy a dual place 'in the wilderness'. Similarly, the Chinese garden, which is based on polar opposites, also tries to mediate between quite contradictory aspects – the extravagant pleasure palace with its harem and the Confucian farmhouse with its simple vegetable plot, the Taoist Abode of the Immortals and the stage-set for boating parties. It is the same kind of 'liminal zone' Leach is describing, and it also serves many of the mediating roles which other ritualistic places also serve, such as church, graveyard and shrine. All these religious or spiritual places seek to break across everyday social time, disrupting the normal flow of events with an abnormal, sacred event. The sacred event must of course take place in real time and space, but it marks a shift in status which is instantaneous and timeless and may even be experienced as being without duration. In a convincing Chinese garden as much as in an effective play, one's sense of discrete, ordered intervals is transcended by a new order – that of the events themselves.

But can the garden really be considered a 'liminal zone' in the greater sense of looking to both the spiritual and earthly realms at once? In support of this idea, one recent scholar, Nelson Wu, who has lived in a Chinese garden and built himself one in Connecticut, has come to similar conclusions from quite a different perspective. In his book *Chinese and Indian Architecture* he notes the current patterns of square and circle which show up in many different Chinese contexts – as plans for buildings, as mandalas – which conventionally signify, respectively, earth and heaven.

> The unnatural shape of the [Chinese] city defines the area within, and at the same time gives meaning to the world outside the square. The equally unnatural order, *li* (social norms and ceremonial rites created by man to govern his behaviour and emotions), is operational only when it has the four walls as its reference of relevancy. That which is 'outside the square (*fang-wai*)' would be outside his concern, once declared Confucius, the master of *li*, and the quotation is significantly in the Taoist classic, *Zhuangzi*, for a different set of values rules that realm. Designed with the basic concept of *tian yuan di fang* (heaven round, earth square), the Chinese mandala, as it radiates from the centre, is a series of alternating and concentric circles and squares beginning and ending with the square (order and knowledge of man), or the circle (chaos and truth of Nature).... The realm outside each square but inside the next larger circle may be conveniently called *tian ren zhi jian,* between Heaven and Man. In this eternally negative space, between reason and untarnished emotion, between the correctness of the straight line and the effortlessness of the curve, between the measurable and the romantic infinity, lies the Chinese garden which is between architecture and landscape painting.[10]

ABOVE AND ABOVE, RIGHT *Two 'places apart'. The scholar half hidden in his pavilion and the Taoist walking under the moon both seek to 'refresh the heart' – the ultimate purpose of the Chinese garden.*

To see how the garden shares this dual, magical form with the mandala we might look at one final garden plan, the Liu Yuan in Suzhou. In the southern and eastern parts of the plan we can make out elements of the Confucian organization of a house. Halls and courtyards, organized symmetrically, face south and convey the predictable social order; and yet their order is broken and complicated by the introduction of many tight spaces, too many to understand, by sudden angles, cut-out vistas and, of course, rocks and trees. These areas are not houses, although people could and did live there. Rather they are the 'liminal' transcendent spaces mid-way between not-quite-living and not-quite-heaven. The main structure in the eastern part is called the Hall of the Mandarin Ducks, a metaphor for conjugal felicity which is carried through in its divided, and joined, symmetrical plan. The occurrence of such ritualistic and social meanings and the omnipresence of such metaphors confirms the dualistic interpretation: such places are meant to be read as occupying two worlds at once. But it is really the overall plan which furnishes a

religious argument to be deciphered. For what can we understand from it? Confusion, heterogeneity, spaceless space and timeless time.

After studying this plan, with its circuitous entrance full of tricks, with its rambling accretions around the centre, its complicated mixing of rock piles and streams, after studying various photographs taken in the garden, which are hard to place, the rather obvious point comes into focus that the design is not meant to be comprehensible: units of scale are not repeated enough to form a pattern; it is mostly variation from a non-existent theme; orientations such as centre/edge, before/after are not very meaningful. It is true that the main lake and *ting* are a kind of centre, but this is rather dissipated by another series of miniature centres and the fact that the garden goes on and on interminably, providing a continuously interesting experience. There is no overwhelming 'sense of an ending' as there is at the Palace of Versailles. Although there is a complicated order that can finally be perceived, the Chinese did not lay out their gardens to be conceptualized from above, in a cerebral helicopter, as the French and Italians did. The Chinese garden was to be perceived as a linear sequence, 'the scroll painting you enter in fancy', that seems infinite. As with much religious experience and ritual, the internal boundaries were made vague or ambiguous, time was made to stop and space became limitless. It was clearly more than an aesthetic game of complexity and contradiction; rather it was a compelling alternative, with its own special meanings, to the world of official responsibilities – a place apart that was, and is, magical and free from the cares of men.

GARDENS ACCESSIBLE TO VISITORS

Note: This list, though greatly enlarged from the 1986 edition, is by no means exhaustive. Old gardens in China, or new gardens on the site of old, are constantly being restored and opened to the public. Many traditional buildings such as temples and houses incorporate garden-like elements in their courtyards and surroundings, though they cannot be described as gardens in themselves; some of these are mentioned below. Grateful thanks to Dr Frances Wood for information on the gardens of Shandong. A list of place-names and garden-names in Chinese characters is included at the end to help visitors find their way to the gardens.

NORTH CHINA
PEKING AREA

Beihai Gongyuan (Beihai Park), lying north-east of the Forbidden City in Peking, was first recorded in the tenth century as the site of an Imperial Lodge. More than half its current 165 acres are made of water, part of a series of lakes (Bei hai means 'Northern Sea'; the others, Zhong hai and Nan hai, mean 'Middle' and 'Southern Sea') originally dug in 1179 and still fed by water from the Jade Fountain Spring near the Western Hills. Earth from this excavation also made the first artificial island hill here, which, enlarged and embellished in turn by Kubilai Khan (when it was described by Marco Polo) and in the Ming and finally Qing dynasties, still exists. However, most of what you see today, including the great white dagoba that dominates the island and the wide white marble bridge that connects it to the shore, dates from the garden-building era of the Qianlong Emperor in the eighteenth century. The gardens around the Middle and Southern Seas (Zhong Nan Hai), used now as government offices and residences, are not open to the public, but Beihai is the central public park in Peking. Boats are for hire on the lake and a wooden ferry crosses to and from the island and the five Dragon Pavilions on the northern shore. A number of 'small sceneries', walled garden complexes and religious courtyard buildings scattered around the lake and on the island, are also open at certain times, including the **Tranquil Heart Studio (Jing Xin Zhai)**, once the residence of the Qing Crown Princes. Renovated in 1982, it is considered one of the best surviving examples of a walled pre-revolutionary Peking garden. The area called Hao Pu Jian ('Between the Hao and Pu Rivers'; an allusion to the Taoist classic *Zhuangzi*), approached by a bridge over a lotus pond, is quite charming and very restrained in style, despite its late date. Some of the 'false mountains' in Beihai are said to have been made of rocks salvaged from the Gen Yue (see pages 67–8), the Song Imperial Garden destroyed in 1126. Also worth seeing are: the beautiful white-barked pines around the Yuan-dynasty carved stone in the Round City, near the entrance to the marble bridge; the statue, on the island, of an Immortal holding up a bowl to collect dew (an idea dating back to the Han Emperor Wudi's search for an elixir of Immortality) and the splendid glazed-tile nine dragon screen. Qionghua Island in Spring was traditionally known as one of the 'Eight Best Sights' of old Peking.

Ning Shou Yuan (Garden of the Qianlong Emperor; Garden of Tranquil Longevity), one of three gardens still existing within the Forbidden City, was built when, out of filial piety, the Emperor decided he should rule no longer than the sixty-one years of his grandfather Kangxi. Thus the garden, a series of five courts arranged along a north-south axis – like a Peking courtyard house in the north-east corner of the palace – was made (and most of its pavilions named) in anticipation of the Emperor's retirement. Two of the courts are simply planted with trees in paved ground. The others are laid out with rockeries, pavilions and *lang* walkways in the

Pavilion on Mochou lake, Nanking.

complex but lyrical style of south-east China, much freer and more informal than the rest of the palace. It includes, in the first courtyard, a 'nine-turns stream', engraved on the floor of the Xi Shang pavilion. The garden was restored and opened to the public in 1982.

Yu Hua Yuan (Imperial Palace Garden), backed up against the high red walls of the Forbidden City, and just within its Northern Gate, is symmetrically arranged around the courtyard of its central 'Hall of Divine Tranquillity'. It forms the last, and most intimate, of the great courtyards that lie on the main north-south axis of the palace – a place for the Emperor to relax after official engagements. Its layout reflects the measured grandeur of court architecture: each pavilion on the left balanced by one on the right; exotic rocks displayed on marble plinths; peonies and chrysanthemums planted in rectangular marble beds. But the formality is balanced by elaborate rockeries and many ancient juniper trees, whose magnificent contorted trunks hold up a high canopy of needles to shade the pavements below. Pebble-paths restored in the late Qing include a bicycle and motor car as designs among traditional vases and pavilions. Reginald Johnston, Scottish tutor to the last Chinese Emperor Puyi once lived in the low house on the western side of this garden.

Zhongshan Gongyuan (Zhongshan Park), named after Dr Sun Yat-sen (Sun Zhongshan), is a public garden laid out behind the great western wall of Tian An Men Gate – the outer entrance to the Forbidden City. Further walls inside divide it into several parts, including one planted with juniper trees that overlooks the high walls and moat of the palace. The first part, mostly flat and divided into rectangular compartments by straight *lang* walkways, is famous for its collection of elaborately bred goldfish displayed during the warmer months in big ceramic water butts.

Diao Yu Tai (Angler's Rest), restored in 1983, is an old walled garden-within-a-garden arranged with rocks, bridges, pavilions and covered walkways around a small, irregular lake. Now included in the grounds of the state guest house, it is a good example of an old Peking garden, somewhat stiffer and more elaborate in feeling than the better-known gardens of south-eastern China.

Gong Wang Fu (Prince Gong's Garden): the garden of Prince Gong's Mansion is located in the north-western part of the old city of Peking, in the area known as the Rear Lakes (Hou Hai). It is said to have been the site of the Jesuit missionary Matteo Ricci's residence when he lived in Peking. Originally it was the property of the Qianlong Emperor's favourite Heshen (1750–99), but after his disgrace and death it became imperial property. In 1851 it came into the possession of Prince Gong (1833–98), who signed the Treaty of Tientsin (Tianjin) in 1860 after the sack of the Old Summer Palace (Yuan Ming Yuan). It was the childhood home of Prince Gong's grandson Pu Ru (Pu Xinyu, 1896–1963), a noted painter who ended his days in Taiwan after 1949; he was the last princely owner of the mansion and garden. It then became the site of a Catholic seminary which was later incorporated into Peking Normal University, and parts of the site are now occupied by the Peking Conservatoire. The garden includes a curious miniaturized 'Great Wall' with a 'Jesuit Baroque' gateway, apparently dating from the time of Heshen. There is an elaborate theatre building in the eastern part of the garden. As a whole, the garden is a good example of the northern style, with its rather imposing architecture. Sunk into the top of the 'artificial mountain' known as Dripping Verdure Cliff (Dicuiyan) are two large pottery vessels to hold the water which would be poured down the rock-face by servants concealed behind it, for the enjoyment of the owner and his guests during the hot summers. The garden is said to have influenced Cao Xueqin's description of Prospect Garden in the great eighteenth-century novel *Dream of the Red Chamber* (*The Story of the Stone*).

Bamboo Garden (Zhu Yuan): now a hotel incorporating a good Sichuan restaurant, this garden residence was occupied from 1956 to 1975 by Kang Sheng (1898–1975), the notorious head of China's secret police, who had an incongruous passion for traditional culture and fine antiques. The ghosts of Kang Sheng's victims, who tormented him on his death-bed, appear to have been laid to rest since the end of the Cultural Revolution, and the garden environment is very pleasant. Reputedly the hotel still belongs to the Ministry of Public Security.

Song Qingling Guju (Song Qingling's Former Residence): After 1949, this was the official residence of Song Qingling (1892–1981), the widow of Sun Yat-sen. In the same area as Prince Gong's Mansion, it was the site of another former Qing princely residence, and traces of the original garden are still visible in the grounds.

Da Guan Yuan ('Prospect Garden'): An entirely modern (1980s) construction, this is a reproduction of the famous fictional garden described in the great eighteenth-century novel *The Dream of the Red Chamber* (*The Story of the Stone*). The site itself is of no historical relevance, and in any case the author, Cao Xueqin (1715?–63), almost certainly envisaged his fictional garden as being somewhere other than Peking, but it may be of interest as a reconstruction of eighteenth-century garden design.

The campus of Peking University, formerly that of Yen-ching University, occupies the sites of two famous Ming dynasty gardens, one of them the **Dipper Garden (Shao Yuan)** which belonged to the calligrapher, rock-collector and connoisseur Mi Wanzhong (1570–1628). Something of the layout of the Dipper Garden's lake may survive in the configuration of the existing Nameless Lake (Wuming hu). The district of Haidian, in which Peking University lies, was the site of many famous late-Ming gardens, mostly belonging to relatives of the imperial family.

Yi He Yuan, known to most foreigners as the **Summer Palace**, was the site of a monastery during the Yuan dynasty, then massively redeveloped under the Qianlong Emperor in the 1760s. Marshy ground, often flooded by the Jade Fountain Stream on its way into Peking, was dredged to increase the hill that overlooks it, and the resulting lake was used for naval training battles. Soon, with his mother's sixtieth birthday as an excuse, the Emperor had halls, courts and pavilions built up the slopes of this 'Longevity Hill', while below, causeways divided the water in imitation of Hangzhou's West Lake. Qianlong's beautiful marble bridges and the little walled **Xie Qu Yuan (Garden of Harmonious Interest)**, inspired by the Jichang garden in Wuxi, also still exist. Partly burnt and looted by the British and French in 1860, the Summer Palace decayed still further through neglect and local pilfering until the Empress Cixi had it rebuilt to celebrate her own sixtieth birthday in 1887, adding, among other curiosities, the great theatre stage and marble boat. It is now a public park. Boats are for hire on the lake and shops and restaurants have been set up in some of the lower courtyards. Most visitors crowd into the more accessible courts and along the famous mile-long *lang* walkway – its beams painted with 1,400 decorative landscapes and gardens – that runs along the foot of Longevity Hill. But since the garden covers more than three square kilometres (four-fifths of it water) there are also many quiet walks and views for those with time and energy.

Ruins of the **Yuan Ming Yuan (Garden of Perfect Brightness; Old Summer Palace)**. For many years some of the most agreeable public places for foreigners based in Peking have been the ruins of the European gardens built in 1785, in the north-east corner of the old summer palace Yuan Ming Yuan (the Garden of Perfect Brightness) near Peking, by the Jesuit father Benoit for the Qianlong Emperor. These ruins, with the just discernible outlines of lakes and islands among the present-day fields, are all that remain of the garden that once, through the letters of Father Attiret, was the talk of Europe. At a conference in 1980, however, Chinese architects and planners called for the garden to be reconstructed as a national park, although this has not yet happened. Contemporary paintings and engravings still exist which could make this possible. Some of the best of these, a series of large watercolours commissioned by the Qianlong Emperor on completion of the garden, can be seen on request in the Bibliothèque Nationale, Paris.

Chengde (formerly known as Jehol or Rehe), lies some 150 miles north-west of Peking, on a river sometimes called Rehe, or 'Hot River', because hot springs keep it ice-free even in the northern winters. Encircled by hills, it is a site of exceptional natural beauty which the Kangxi Emperor first used as a stopover on the court's annual progress to the Manchu homelands. Commissioned by Kangxi in 1703, the **Bi Shu Shan Zhuang (Imperial Summer Retreat)** was celebrated first by him, in a series of poems and woodcuts of its thirty-six *jing* or 'scenes', then later by his grandson the Qianlong Emperor who repaired them and added thirty-six more. For political reasons he also built twelve great temple complexes on the eastern slopes, of which seven survive. The Emperors spent so much time at Chengde that it became of necessity an important political centre, and large complexes of administrative buildings were built to the south of the park. This park, the largest in China, is encircled by a wall that winds up and down among the low hills that make up four fifths of its total. Pine and willow-planted lowlands, lakes and causeways – one modelled on the Su Causeway at Hangzhou, each with its own series of pavilions – make up the rest. Among them, shimmering on the lake, is Green Lotus Island with its Hall of Misty Rain and the strange, naturally shaped Pestle Rock 'borrowed' into the view from the horizon behind. Chengde, less crowded because less accessible than the Yi He Yuan (Summer Palace) or Hangzhou's West Lake, is – especially in autumn when the lakes are misty, the sky blue and the hills russet gold – one of the great sights of China.

SHANDONG PROVINCE

Jinan

Great Brightness Lake (Da Ming Hu), first mentioned by a famous geographer in the Northern Wei, lies in what is now the centre of Jinan city and is fed by seventy-two famous springs. In 1072, a celebrated writer and governor of the city, Zeng Gong (1019–83), dammed the edges to make use of the water, and since then the temple and garden complexes built around its shores, and poems written in its praise, have made it almost as famous as Hangzhou's West Lake. On the north bank of the lake is the North Pole Pavilion, flanked on the east by the memorial temple to Zeng Gong. To the west is the memorial temple of Tie Xuan (1366–1402), a Ming dynasty governor who defended the city against a usurper. It contains many stone tablets of calligraphy let into the walls; one, written by Tie in his own hand, reads: 'Around the gleaming lotus flowers – green green willows; around the softly singing lake – the verdant city.' The park which now encloses Great Brightness Lake incorporates several smaller gardens, including the **Xia Yuan (Distant Garden)** and the **Xiao Cang Lang (Lesser Blue Waves**, referring to the Blue Waves Pavilion in Suzhou); the Xiao Cang Lang was constructed in the late eighteenth century.

The city of Jinan is celebrated for its springs. The **Baotu Quan (Fountain Spring)**, in the south-west corner of the old city, is said to be not just the finest of the seventy-two, but 'the first spring under heaven'. Its garden setting is focused on the water of the spring.

Qufu

The **Kong Fu (Confucian Mansion)** was the residence of the descendants of the sage Confucius until the 77th Yansheng (Extension of the Sage) Duke followed the Nationalist government in 1940 to Chongqing and then Taiwan. At the rear of the mansion complex is the **Tie Shan Yuan (Iron Hill Garden)**, first constructed in 1503 and renovated in the eighteenth century when a daughter of the Qianlong Emperor was married to the 72nd Yansheng Duke. A typical northern-style garden, it includes a 400-year-old cypress tree.

Zibo

On the outskirts of the ancient city of Zibo is **Pu Songling Guju (Pu Songling's House)**, a good example of Qing dynasty vernacular architecture with courtyard gardens. Pu Songling (1640–1715), who repeatedly failed the civil service examinations during his life, became famous after his death as a writer, principally of supernatural short stories, many of which have been translated into English and other European languages.

EAST CHINA

NANKING (NANJING)

Zhan Yuan (Outlook Garden). A garden was first made on this site by Xu Da, a general famous in China for having helped bring the Ming dynasty to power. During the Qing, the Qianlong Emperor twice visited it, and sealed its fame by writing, in his own hand, an inscription above the southern entrance. Restored and rebuilt twice more, the house was used by one of the Taiping leaders and is now a museum of that rebellion. Leading directly from it to the west, the garden is made up of two ponds, each with an interesting rockery, connected by a stream on the west and a *lang* gallery on the east. The garden was repaired after the founding of the People's Republic in 1949, and the northern mountain is in the traditional style. The innovative, finely conceived and crafted southern mountain, however, was made only in 1960. Flat rocks are used to form natural-looking terraces on the pond, while the 'mountain' itself rises gradually to form a shallow cavern, from whose rocky creviced heights water falls precipitously to the pool below. A fine Taihu Rock, the 'Immortals' Peak', half visible from the street through a window in the entrance hall, is reputed to have belonged to the Song Emperor Huizong.

Xu Yuan (Warm Garden): Like the Zhan Yuan, this was the site of a garden belonging to one of the generals who helped bring the Ming dynasty to power. In the Qing period, it was an official government residence, but was at one point occupied by the leader of the Taiping Rebellion.

Mochou Hu (Mochou Lake) is named after a Lady Mochou said to have died tragically here in the Southern Qi dynasty. A pretty pool surrounded by willows, it has been famous as 'the first lake of Nanking' since the Song and Yuan dynasties. Two southern courtyards with some fine old halls and views through successive doorways to the lake have been renovated, and a modern pavilion added to encourage walks along the shore. But as a whole it now lacks the complex intensity of great gardens – something which the addition, in the middle of the fishpond, of a white statuette of Lady Mochou in antique dress, does little to improve. In the vicinity of this lake, in the late Ming, was a garden belonging to the father of Ruan Dacheng (*c.* 1587–1646), a patron of the garden designer and theorist Ji Cheng (b. 1582); the garden may well have been designed by Ji Cheng himself, but no trace of it remains.

WUXI

Ji Chang Yuan (Garden for Lodging One's Expansive Feelings), a walled Ming dynasty garden within a larger park, is famous for its borrowed views, to Mount Hui with its pagoda in the south and to Mount Xi in the west, where the garden rockwork is masterfully arranged to look as if it were a natural rocky spur of the hill. However, Zhang Shi, the rock-craftsman who built it during the reign of the Kangxi Emperor, clearly did not anticipate the electric-light standards in modern Chinese baroque which have been added to his work, or that the garden's Embroidery Ripple Lake would one day reflect not only the pagoda on Mount Hui, but also two large telephone poles. Nevertheless it is a wonderful garden full of delicate juxtapositions, and with a lyrical feel quite different from the more formal Peking garden built in its imitation, the Xie Qu Yuan, commissioned for the Summer Palace by the Qianlong Emperor (see pages 80–81). The garden originally belonged to the Qin family, which since at least the Song dynasty has produced a number of noted scholar-officials and literary men; the family's history has been recounted by a scion, the journalist Frank Ching (Qin), in his book *Ancestors*, which includes some information about the garden.

Li Yuan (Li Garden) is a twentieth-century garden. Begun in the 1920s, it was increased after the founding of the People's Republic. It runs along the shore of a wide stretch of water attached to the great Lake Tai and is in two parts, connected by a long *lang* walkway with open views across the garden and lake on one side and, on the other, of inland fields screened by decorative windows. Close up, some parts are coarsely conceived and constructed, and the great rockery topped off with a concrete pavilion seems excessive and exaggerated. But the 1950s pavilions of the Four Seasons, set squarely round a stiff pool on the edge of the lake, are full of grace. Beyond them, borrowed views of water, sky and hills with pavilions silhouetted in the foreground, create some magical effects. (See the photographs on pages 177 and 180).

Mei Yuan (Prunus Garden): like the Li Yuan, this lies beside Lake Tai. It is a private garden dating from the late nineteenth or early twentieth century. As the name suggests, it is noted for its prunus blossom in the early spring, and also has the strongly scented autumn-flowering osmanthus, but otherwise it is of no particular distinction.

YANGZHOU

Slender West Lake (Shou Xihu) was gradually developed in Sui and Tang times from a right-angled bend of the old city moat. During the Qing, given the impetus of several visits of the Qianlong Emperor, twenty-five compositions of halls, pavilions, villas and bridges were added round its shores. The most striking image today is of the 'Five Pavilion Bridge', its slender scarlet pillars holding up a symmetrical cluster of imperial yellow roofs above its massive central arch. To its south, a white dagoba, smaller than the one at Beihai in Peking, is said to have been temporarily constructed overnight of salt, a major local product, to please Qianlong, then built of more lasting materials for his next visit. Around the shores of Slender West Lake were built a number of private gardens, including the **Xu Yuan (Xu Garden)**.

The **Ge Yuan** (which could be interpreted as **Individual** or **Isolated Garden**) was created during the early nineteenth century for a successful salt merchant (Yangzhou was an important centre of the salt industry). The name, written with a character which, if duplicated, forms the word for 'bamboo', refers to the bamboos widely planted by its second owner and suggests his aspirations towards cultured status: an early literatus said of bamboo, 'How can one live one day without these gentlemen?' Bamboo, with accompanying 'stone bamboo shoots' (a type of tall slender greenish rock), is a feature of the garden. It is also famous for its four miniature mountains built in different types of rock and planted with different plants, to symbolize the four seasons.

The **He Yuan (He Garden)** is another salt merchant's garden, of later date than the Ge Yuan; it is the largest surviving private garden in Yangzhou. There was another garden on the site in the eighteenth century, but the present garden dates from the last quarter of the nineteenth century. It belonged to the He family, hence the name by which it is usually known; its formal name is **Jixiao Shan Zhuang**, literally 'the mountain villa in which one gives vent to a roar', referring to a type of vocalization connected with Taoist breathing practices. The proportion of buildings to rocks, water and plants is very high, and there are even some touches of Western-style decoration. The garden is very much a product of the late nineteenth century. An interesting feature is the small theatre building, set in the pond in the main courtyard where it would be visible from the buildings around, both to male guests and to the women of the family (who would be concealed behind the shutters of the upper storey). Both Ge yuan and He yuan are regarded as characteristic of Yangzhou – less soft and lyrical than those of Suzhou – but strong conceptually and well worth restoring to their former grace.

Ping Shan Tang (Hall Level with the Mountains), adjoining **Daming si (Temple of Great Enlightenment)**, is on the site of a garden residence constructed in 1048 by the great Song dynasty scholar official Ouyang Xiu (1007–72). He named it, during a period of official disfavour, ostensibly in reference to its elevated site and its spatial relation to the hills around Yangzhou, but more subtly to imply that he was still the equal of those officials in high favour at court.

Pian Shi Shan Fang (Sliver of Rock Mountain Cottage): This garden is said to have been designed and occupied by the distinguished late-Ming/early-Qing painter Zhu Da (Bada shanren, *c.* 1626–*c.* 1705). However, the garden to be seen today appears to be an entirely modern reconstruction.

Yangzhou is also the site of a number of smaller once private gardens, which may or may not be accessible to the public depending on circumstances.

SUZHOU

A warm climate, abundant water, long prosperity and a large population of scholars and artists have made Suzhou the most famous city in the country for gardens. Many have disappeared but some of the best known have been restored and are open to the public. Nearly always crowded, their intricate complexity is confusing on a short tour, but for the first, sunny, empty half hour of a spring morning they can still be the most magical places in China.

Shi Zi Lin (Lion Grove) was commissioned in about 1336 by a Buddhist monk Tianru, as the northern section of the Temple of Bodhi Orthodoxy. It is famous for its rockery and, since the words for 'teacher' and 'lion' are homophones in Chinese, its name is a pun on the lion-shaped stones in the garden and a reference to the famous, naturally formed Lion Cliff at Tianmu mountain, where the monk's master once had a retreat. More than ten well-known artists are said to have worked on the design; although this is probably only a legend, one of them, Ni Zan, made a scroll of it in about 1380. However the garden today, with its vast pile of contorted rocks spilling out around the lake in peaks and caverns, bears no resemblance to the rather austere, rocky-edged courtyard he recorded. A long *lang* walkway runs around much of the garden, with many engraved stone tablets and intricate 'flower windows' let into its walls. Twentieth-century additions include a concrete and wood 'stone boat' in the north-east corner of the lake, pavilion windows in brightly coloured stained glass, and bare-bulb electric lighting. All seem to me unfortunate. It was last privately owned by the family of the American architect I.M. Pei.

Wang Shi Yuan, or the **Garden of the Master of the Fishing Nets,** was first built in 1440 by an official from Yangzhou, then abandoned after his death and rebuilt in 1770. Famous during the eighteenth century for its peonies, it changed hands frequently in the nineteenth and was finally restored in the 1940s by the Wang family. One of the most unspoiled of the old gardens, it covers only one acre and is unique in having open to the public the large, shadowy halls of the house to which it was attached (the house was still occupied when taken over by the city in 1958). The small Courtyard of the Late Spring Abode (Dian Chun Yi) in the north-west corner of the garden inspired the design for the Astor Court, a Chinese courtyard built in 1981 to complement Chinese works of art in the Metropolitan Museum in New York. (See also footnote 1 to the chapter on Meanings of the Chinese Garden.)

The **Zhuo Zheng Yuan,** usually known as the Garden of the Unsuccessful Politician but better translated as **Garden of the Artless Administrator** (in allusion to the words of a discontented third-century AD scholar-official who claimed that gardening was the only form of administration suited to the artless), was the site of a Confucian scholar's house in the Tang dynasty and a monastery under the Yuan before being rebuilt by a successful official, Wang Xianchen, in the Ming. His son lost it gambling and it changed hands many times before it was acquired by Wang Yongkang, son-in-law of the notorious General Wu Sangui who let the Manchu conquerors overcome China. Often sold, repaired, ruined and repaired again, it was used as an office by a revolutionary general during the Taiping rebellion. Sirén photographed it in the 1930s, very overgrown and with its ponds silted up, but still extraordinarily evocative. Today, restored again, it is in three parts, divided by walls, which together make it one of the largest of the old gardens. The eastern part, added after the establishment of the People's Republic, has been laid out in a modern, Western-inspired style with grassy lawns, few pavilions and pleasant plantings of trees on mounded hills. The central and western sections are more intricate and interesting. Composed around a winding lake which elongates into long, stream-like fingers, it is admired for an atmosphere 'like the water villages of Jiangnan', the region south of the Yangtze in which Suzhou lies. It is celebrated for its association with the great Ming painter Wen Zhengming, who painted and wrote about it, and with the poet-playwright Yuan Mei, who often visited it in the eighteenth century.

Ou Yuan, meaning Couple's or Coupled Garden, is the name of a garden to the east and west of an old city residence. First built by Lu Jin, a District Magistrate in the early Qing dynasty, the garden acquired its present form under Shen Bingcheng, an Intendant of Circuit in the late Qing. The western is divided by a studio building into two courtyards, the one in front with an unusual rockery. The eastern garden is one of the most interesting in the city. Inside its entrance a small courtyard leads into a relatively large, undeveloped space bounded by a white wall with several ornamental *lou chuang* windows cut into it and a fine moon gate framing the garden beyond. Covered *lang* walkways lead away inside this gate, following the wall, then breaking free to zig-zag across the garden. The main hall, two-storied and double-eaved, is hidden from the entrance by a high and very articulated artificial 'mountain' of yellow rocks, which falls steeply on its eastern flank to a narrow reflecting pool. It is much admired for its craftsmanship. Winding steps are cut along this miniaturized cliff to give visitors the sensation of walking through a profound valley. A recently paved path here is worked in a checkerboard, with the checks slanting off to the right, pulling the eye towards the succession of decorative windows in the enclosing wall. Along the eastern boundary a double *lou* allows views of the busy canal and city outside the garden. Not usually visited by tourist groups, the Ou Yuan is a little run-down but, in its delicate manipulation of space and light and its many peaceful and hidden corners, it is a classic of Suzhou garden art.

The Liu Yuan, or **Garden to Linger In**, dates from the Ming dynasty, when it was laid out by an official named Xu Shitai. It acquired its present name under a later owner, Liu Yongfeng, whose 'Liu' sounds like the Chinese word for 'keep' or 'stay', a reference to the survival of the garden through the Taiping rebellion. Arranged in four main sections, it is the largest of the remaining gardens (nearly ten acres), and is famous for its fine grill-work windows, its elaborate pavings, its complex labyrinth of halls and corridors, its 300-odd pieces of calligraphy let into the walls on engraved stone tablets, and the 'cloud crowned peak', a Taihu stone some fifteen feet high which is the focus of its own courtyard.

The Xi Yuan, or Western Garden, was laid out to the west of the Liu Yuan at the same time and by the same man. Later it became a monastery, and still contains a Hall of Ten Thousand Buddhas. Though pleasant, the garden is not particularly remarkable, and is most remembered for its large, two-storied pavilion, reached by a zig-zag bridge from either side of its central lake.

The Cang Lang Ting or **Blue Waves Pavilion** still has the same layout as it did in the Song dynasty when, in 1044, it was designed on the site of a tenth-century house. Its name, given by the poet Su Shunqin (Su Zimei), comes from a line in the *Chu Ci* poems: 'If the water of the Cang Lang river is clean, I wash the ribbons of my official hat in it; if it is dirty I wash my feet'. Destroyed in the Taiping rebellion, it was rebuilt in 1927, when it became part of an art college, and next to it on the canal is a somewhat coarse building in a classical Western style. The actual Cang Lang Ting is now made of concrete, but the garden has some beautiful courts and bamboos, and some dramatic decorative windows which look to Europeans somewhat like Art Nouveau. Effects of light are very well used, particularly in the series of three rectangular rooms crosslit by latticework windows and joined at the corners to make a long, diagonal 'corridor'. Sunlight, streaming through tall plantings of bamboo on either side, here turns the rooms almost green, like a cavern underwater. A double-storied *lou* opens on views of hills beyond the city. The Cang Lang Ting is also famous for the way it 'borrows' the canal outside the entrance gate into the garden. Since a walkway with windows through it runs on both sides ofthe perimeter wall, visitors cannot tell exactly where the garden starts and stops, and it feels much larger than it is.

The Yi Yuan or **Garden of Pleasure** is relatively recent, built in the late Qing by a rich official on the site of a garden once owned by the Ming calligrapher Wu Kuan. It is celebrated for its clever adaptations of features of other, older gardens, and though small it is an intricate, delicate and delightful garden.

The Yi Pu (Garden of Arts) is one of the most attractive of Suzhou's smaller gardens. Its site was once owned by Wen Zhenmeng (1574–1636), a great-grandson of the sixteenth-century painter, poet and calligrapher Wen Zhengming. Wen Zhenming's brother Zhenheng (1585–1645) is known as a connoisseur of elegant living whose *Treatise on Superfluous Things* (*Zhangwu Zhi*) expresses his views on garden style, among many other topics. One of the pavilions in the garden, the Ru Yu Ting (Hatchling Fish Pavilion), is believed to date back to the Ming dynasty; the large rockery adjoining it may also be Ming in origin. The building facing the rockery on the other side of the pond is now a teahouse.

The **Huan Xiu Shan Zhuang (Mountain Villa Surrounded by Flourishing Greenery)** is now part of the Suzhou Embroidery Institute. Although easily accessible to tourists, it is not open to the general public as a garden, so despite its small size it is usually quite uncrowded. It dates from the mid-Qing dynasty. The very remarkable artificial mountain is the work of a well-known eighttenth-century designer, Ge Yuliang. There are so many winding paths and ravines within its small compass that visitors can spend a considerable time exploring it without retracing their steps. Unfortunately the vegetation in the garden has been quite neglected and the well-established trees and shrubs which once grew on the 'mountain' have been allowed to die, leaving it looking much starker than it should.

The **Crane Garden (He Yuan)** houses the Suzhou office of the Chinese People's Political Consultative Conference (CPPCC) and is not officially open to the public. However, Chinese-speaking foreign visitors have been known to charm their way in. It is a pleasant, small nineteenth- or early twentieth-century private garden, though not worth going out of one's way to visit.

Tiger Mountain (Hu Qiu), outside Suchow, is an ancient and famous site, crowned by a leaning pagoda more than 1,000 years old. Around it are placed some halls and pavilions among natural rocks. There is a fine view of the countryside from the top of this pagoda.

SHANGHAI AREA

The **Yu Yuan** or **Garden to Please** in the old 'Chinese City' of Shanghai, is a large and complex garden, constructed by a retired official, Pan Yunduan, in 1559 for the enjoyment of his elderly father, Pan En. The garden has been frequently damaged and rebuilt since then, but the large artificial mountain constructed of yellow rocks is believed to have survived from the original garden; it was the work of a local rockery designer, Zhang Nanyang. The garden contains a number of dramatic individual rocks, including Yulinglong (Exquisite Jade), which is said to have been left behind by the rock collectors commissioned by the Song emperor Huizong to acquire material for his Gen Yue garden.

The **Gu Yi Yuan** or **Ancient Garden of Elegance** in Nanxiang, a semi-rural town belonging to Shanghai Municipality, dates back to the sixteenth century, when it was owned by a family called Min; it is said to have been designed by the famous bamboo carver Zhu Sansong, many of whose works in bamboo still survive. It has undergone frequent damage and reconstruction; at one point it became the garden of the city god's temple. It was sensitively restored in the 1980's under the guidance of the distinguished garden historian Professor Chen Congzhou.

The **Garden of Autumn Vapours (Qiu Xia Pu)** in Jiading, a county town within Shanghai Municipality, has alternated over the years between private and temple garden. Part of the site was a private garden from the early sixteenth century, until it became a temple garden in the early eighteenth century. This garden was amalgamated with the private garden next door later in the eighteenth century, to form the present site. From its temple garden days the garden retains an impressive theatre stage. It also retains a more regular and geometric layout than a purely private garden would have. It has some very attractive decoratively shaped doorways.

Both the Gu Yi Yuan and the Qiu Xia Pu can be visited from Shanghai by hiring a taxi for the day.

The **Lesser Lotus Manor (Xiao Lian Zhuang)** in Nanxun belonged to a family of entrepreneurs and book collectors, and the garden contains a large early-twentieth-century library building. Some of the garden's buildings show a strong Western influence, but still fit quite attractively into their setting around a very large lotus pond. In summer, with the lotuses in bloom, it is a fine sight. A day trip from Shanghai to this garden can be combined, for those interested, with a visit to the impressive and historic Catholic church at Sheshan.

Now an undistinguished small (by Chinese standards) town within Shanghai municipality, Songjiang was once an important commercial and cultural centre where significant developments in Chinese garden history took place. Its only surviving garden is the **Zui Bai Chi (Drunken Bai Pond)**. The present garden is in two parts, east and west; the east part is the old garden. It is said to have been the residence of the sixteenth–seventeenth century art critic Dong Qichang. His house, with its priceless art collection, was burnt down by a mob outraged by the liberties his son had taken with a young local widow. A name-plaque in Dong's calligraphy survives. A garden was created on the site in the mid-seventeenth century by a local artist, Gu Dashen, who supposedly designed the pond as its main feature to commemorate his work as a government official on irrigation projects. He named it in honour of the Tang poet Bai Juyi, who enjoyed drinking beside the pond of his home in Luoyang (not, as is sometimes said, the earlier, more famous and more bibulous Tang poet Li Bai). It has served since then as the site of an orphanage and a Japanese military brothel, but has now been reconstructed and opened to the public.

HANGZHOU

Xi Hu, the West Lake, surrounded on three sides by a ring of hills is, among all the 'hills and waters' of China, the most loved and copied. Over a recorded history of some 2,000 years the original lake has been so transformed that it is in fact virtually manmade, and part of its extraordinary appeal is the vision it suggests of man and nature in perfect harmony. The two main willow-planted causeways which divide it are named after two great poet-administrators of the city, Bai Juyi and Su Dongpo, who dredged and reinforced the lake to control floods. Later Hangzhou became the capital of the Southern Song and many fine villas were built along its shores. In the lake is the clover-leaf shaped island called 'Three Pools Mirroring the Moon', which is one of the most beautiful places in China.

On the shore of the 'inner' lake, some way from the area most frequented by visitors, is the **Guo Zhuang (Mansion of the Guo Family)**, also known as the **Fenyang Villa (Fenyang Bieshu)** from the supposed birthplace in Shanxi of the Guo clan. Although constructed only in 1904, and ruined and restored since then, it is charming and quite classical in style, with a wonderful borrowed view over the West Lake. It is a favourite spot for local brides to have their photographs taken. It also has a collection of bizarre furniture made of tree- and bamboo-roots.

In the hills around the lake are many famous monasteries, of which the best known is the Ling Yin Temple.

SHAOXING

Shaoxing in Zhejiang province, a couple of hours by train or bus from Hangzhou, is the site of Wang Xizhi's **Lan Ting (Orchid Pavilion)**, where literary men of the fourth century AD gathered to drink wine and compose poems along the banks of a winding stream. The site of the Orchid Pavilion is a short distance outside the city. Everything to be seen there – the winding stream, the pavilions, the Goose Pond whose occupants' sinuous movements inspired Wang Xizhi's calligraphy – is a later reconstruction, but very attractive none the less.

Within the city, the **Shen Yuan (Garden of the Shen Family)** is supposed to be the site where the Song dynasty poet Lu You (1125–1210), a native of Shaoxing, once encountered the wife whom his domineering mother had forced him to divorce. In passionate grief, he inscribed a poem on one wall; some time later, when the former Mrs Lu returned to the garden and saw the poem, she too wrote a poem there, matching the metre and rhymes of Lu You's original. Lu You's poem remains one of his best known and most moving. This garden, which had become quite ruinous in the twentieth century, is one of the few in China where it has been possible to carry out any archaeological investigation. A low earthen mound and a gourd-shaped well of Song date were discovered, and the whole garden has been reconstructed in what appears to be a reasonably authentic Song-dynasty style.

The **Qing Teng Shuwu (Green Vine Studio)** was the residence of two famous Ming literati, first the mad painter and dramatist Xu Wei (1521–93) and later the 'mannerist' painter and illustrator Chen Hongshou (1599–1652). There is a green vine or creeper, though it may not be the one known to Xu Wei, in the tiny courtyard which encloses the little pond called 'Pool of Heaven' from which Xu took his cognomen 'Taoist of the Pool of Heaven' (Tianchi daoren). In another courtyard there is a small rockery structure built up against a wall, bearing the name Self-Possession (or Easy-Going) Cliff (Zizai yan). As a whole, this must be one of the most delightful garden residences in China.

Shaoxing continued to produce literary men, and women, into the nineteenth and twentieth centuries. The **Qiu Jin Guju**, home of the late Qing woman poet and martyr Qiu Jin (1879–1907) and the private school once attended by the writer Lu Xun (1881–1936), the **San Wei Shuwu (Three Flavours Studio)**, both provide good examples of traditional architecture incorporating garden elements.

NINGPO

The prosperous port of Ningpo in Zhejiang province is the location of the **Tian Yi Ge Library**, whose grounds are laid out in traditional garden style. Originally a private library owned by the Fan family, and now the oldest surviving library in China, the Tianyige dates back to 1561. The name, literally 'Heavenly Unity Pavilion', alludes to a saying in the *Yi Jing (Book of Change)*, 'Heavenly Unity gives birth to water'; the invocation of water was obviously essential in connection with a wooden building containing thousands of flammable volumes, before the days of fire extinguishers. The library, which became the model for the four imperial libraries of the Qing dynasty, is of great interest in itself as well as for its surrounding gardens.

SHEXIAN/HUIZHOU

The town of Shexian in southern Anhui province was formerly the administrative seat of Huizhou prefecture, whose merchants were renowned throughout China for their control of the salt trade (many, if not most, of the Yangzhou salt merchants originated from Huizhou) and of most aspects of the publishing industry, including printing, paper-making, and ink production. The region is also famous for its 'three types of carving', in stone, brick and wood, and its inhabitants pride themselves on its distinctive architecture. Historically, the style of Anhui gardens seems also to have been distinct from that of the cultural heartland of Jiangnan. Owing to the region's present poverty, the architecture is generally quite well preserved, but not much survives of the gardens. The **Xin'an Bei Yuan (Xin'an Garden of Stelai)**, a collection of calligraphic inscriptions on stone preserved in a garden setting just outside Shexian town (near the Tai Bai Lou), may give some idea of a traditional garden of the region; Xin'an is an old name for the area. Some surrounding villages also preserve (or are restoring) their traditional 'water mouth gardens' (*shui kou yuan*), a kind of small public garden set around a stream where it passes out of the village; these were often constructed as a donation to the community by successful local merchants. An example can be seen at Tangmo village. Shexian is readily accessible by train from Shanghai (the journey takes about eight hours), but facilities for foreign tourists are extremely limited and a visit should probably not be attempted without a Chinese speaker in the party. Many tourists, however, pass through Tunxi airport on their way to Huang Shan (Mount Huang); the decorative rocks outside the terminal building are said to have come from the Zuo Yin Yuan (Garden of Sitting in Reclusion), in nearby Xiuning, which belonged to the late Ming publisher, playwright and chess-champion Wang Tingna. Little trace of the garden survives on the ground, but a magnificent woodblock-printed scroll, entitled *Huan Cuitang Yuan Jing Tu (Garden Scenes from the Hall Surrounded by Jade)*, published by its proud owner in about 1608, can still be enjoyed in facsimile (the original was lost in the Cultural Revolution).

WEST CHINA
XI'AN

Hua Qing Spring (Hua Qing Chi), with paths and pavilions above it, lies on the northern slopes of Mount Li some fifteen miles from Xi'an. The hot spring here was already the site of a royal 'short-stay villa' for the First Emperor of Qin, who united China in 221 BC and whose capital, Chang'an, was near present-day Xi'an. The garden's great fame, however, dates from the Tang Emperor Ming Huang, who developed it into a great palace complex where he spent the winters with the Lady Yang Guifei, one of the Four Famous Beauties of Chinese history. Later in the dynasty Bai Juyi's 'Song of Everlasting Regret', which celebrates their story, made every educated Chinese familiar with the names and settings of the garden, though the places themselves were destroyed in the An Lushan rebellion which forced Ming Huang to have his Lady executed. A monastery during the Five Dynasties, and reconstructed under the Qing, it was last privately used by Chiang Kai-shek, captured in a pavilion here by the local warlord during the Xi'an Incident of 1936. Foundations of the Tang palace still exist, and it is possible to bathe in the relaxing mineral waters among courts and pavilions bearing the names of those in which Lady Yang captivated the Emperor. But the buildings themselves date mostly from the 1950s and the garden with its lake and hillside pavilions by rocky streams covers only a small part of the original.

CHENGDU

The **Du Fu Caotang (Thatched Hut of Du Fu)** is a northern Song reconstruction of the scholar's hut once lived in by Du Fu, the greatest Tang poet (perhaps the greatest of all Chinese poets). During each successive dynasty since the Tang, the garden has been restored and added to, the last time in 1949, and it now includes an elongated pond with a number of pavilions and halls set among trees. The Wu Hou Ci (Memorial Hall of the Martial Marquis) commemorates a famous statesman of the Three Kingdoms period, Zhuge Liang (181–234), ennobled as the Martial Marquis, and his master Liu Bei, the Zhaolie Emperor (161–223). Grey-roofed, pink-washed stucco walls wind along a serpentine path which joins the two Memorials and gardens, and above them soar huge tree bamboos which cast shadows on the walls below.

The **Qing Cheng Shan (Green Fortification Mountain)**, outside Chengdu, is the site of a Taoist temple complex, which provides a good example of rural temple architecture forming a 'designed landscape'.

SOUTH CHINA
CANTON (GUANGZHOU)

There are three famous old restaurants in this city where patrons eat in innumerable courtyards and halls of old gardens, all rebuilt here into complex new arrangements. They are exceptionally enjoyable places to eat in, but as gardens *per se,* somewhat on the vulgar side.

The **Orchid Nursery (Lan Pu)**, near the China Hotel and more or less opposite the Canton Trade Fair site, is a modern but very pleasant public garden specializing in the raising of orchids (cymbidium and others). With its sub-tropical vegetation, it gives the visitor a strong impression of the differences in garden styles created by different climates. It also contains a small Suzhou-style garden originally designed for a garden exhibition in Germany.

Of the 'four famous gardens of Guangdong' (which are not, in fact, famous at all outside their home province), perhaps the most charming is the **Yu Yin Shan Fang (Mountain Cottage of Abundant Shade)** in Panyu. Dating from 1866, this garden shows an intriguing mixture of Chinese and European influence, which is characteristic of Cantonese gardens. It has a built-in bamboo peacock cage, which unfortunately, or perhaps fortunately, no longer houses a peacock. Panyu can be reached by ferry from Hong Kong. The **Qing Hui Yuan (Garden of Pure Splendour)** in Shunde, on the other hand, has more concrete than vegetation, and is not worth visiting. The **Liang Yuan (Garden of the Liang Family)**, also known as the **Qun Xing Caotang (Thatched Hall of Assembled Stars)**, in Foshan is now the headquarters of the Guangdong Penjing (Bonsai) Association, with many fine bonsai on display; the visitor should also take in the Foshan Ancestral Temple with its exuberant ceramic decoration, and can also visit the ceramic factories for which the town is famous. Of the 'four famous gardens', these three are all on the western side of the Pearl River Delta. The fourth, in the now flourishing manufacturing town of Dongguan on the eastern side, is the **Ke Yuan (Could-be Garden)**. Dating from the third quarter of the nineteenth century, this garden has the typical brick buildings of the region, again showing some European influence but also including an imitation of the defensive towers common in rural Guangdong villages; here it provided the garden's owners with both an elevated viewpoint and a cool sleeping place during the sultry summer months. Part of the garden adjoins what is now a rather squalid pond but at one time was evidently the local equivalent of the English village duck-pond. The garden is attractive and worth a visit; Dongguan can easily be reached by bus either from Hong Kong or from Canton.

HONG KONG

After the demolition in the 1990s of the 'Kowloon Walled City', which was not in fact walled within living memory, but had an anomalous legal status, remaining nominally under Chinese jurisdiction within the British colony, a public park was created on the site. This includes a well-designed 'traditional' Chinese garden, owing more to the style of Jiangnan than Guangdong. Some archaeological remains of the old walled city can also be seen nearby.

MACAO

The ninteenth-century **Lou Lim-Ieok Garden** in Macao, once owned by the Lou family and now a public park, has some similarity to the 'four famous gardens of Guangdong' in its combination of southern Chinese and European influences, though it tends much more towards the Portuguese style. A pleasant oasis within the increasingly bustling city of Macao, it is used for performances during the Macao Music Festival. Like other southern gardens, it provides a reminder of how much garden style can be influenced by climate and vegetation.

GARDEN NAMES IN CHINESE

ENGLISH TRANSLATION	ROMANIZATION	CHARACTERS
NORTH CHINA		
PEKING (BEIJING)		北京
Beihai [North Lake] Park	Beihai Gongyuan	北海公园
Tranquil Heart Studio	Jing Xin Zhai	静心斋
Between the Hao and Pu [Rivers]	Hao Pu Jian	濠濮间
The Garden of the Qianlong Emperor (Garden of Tranquil Longevity)	Ning Shou Yuan	宁寿园
The Imperial Palace Garden	Yu Hua Yuan	御花园
Zhongshan Park	Zhongshan Gongyuan	中山公园
Angler's Rest	Diao Yu Tai	钓鱼台
Prince Gong's Garden	Gong Wang Fu	恭王府
Bamboo Garden	Zhu Yuan	竹园
Song Qingling's Former Residence	Song Qingling Guju	宋庆龄故居
"Prospect Garden"	Da Guan Yuan	大观园
Dipper Garden	Shao Yuan	勺园
Summer Palace	Yi He Yuan	颐和园
Garden of Harmonious Interest	Xie Qu Yuan	协趣园
Garden of Perfect Brightness ("Old Summer Palace")	Yuan Ming Yuan	圆明园
Chengde		承德
Imperial Summer Retreat	Bi Shu Shan Zhuang	避暑山庄
SHANDONG PROVINCE		山东
Jinan		济南
Great Brightness Lake	Da Ming Hu	大明湖
Distant Garden	Xia Yuan	遐园
Lesser Blue Waves	Xiao Cang Lang	小沧浪
Fountain Spring	Baotu Quan	趵突泉
Qufu		曲阜
Confucian Mansion	Kong Fu	孔府
Iron Hill Garden	Tie Shan Yuan	铁山园
Zibo		淄博
Pu Songling's House	Pu Songling Guju	蒲松龄故居
EAST CHINA		
JIANGSU		江苏
Nanking (Nanjing)		南京
Outlook Garden	Zhan Yuan	瞻园

ENGLISH TRANSLATION	ROMANIZATION	CHARACTERS
Warm Garden	Xu Yuan	煦园
Mochou Lake	Mochou Hu	莫愁湖
Wuxi		无锡
Garden for Lodging One's Expansive Feelings	Ji Chang Yuan	寄畅园
Li Garden	Li Yuan	蠡园
Prunus Garden	Mei Yuan	梅园
Yangzhou		扬州
Slender West Lake	Shou Xi Hu	瘦西湖
Xu Garden	Xu Yuan	徐园
Individual or Isolated Garden	Ge Yuan	个园
He Garden	He Yuan	何园
(Mountain villa in which one gives vent to a roar)	Ji Xiao Shan Zhuang	寄啸山庄
Hall Level with the Mountains	Ping Shan Tang	平山堂
Temple of Great Enlightenment	Da Ming Si	大明寺
Sliver of Rock Mountain Cottage	Pian Shi Shan Fang	片石山房
Suzhou		苏州
Lion Grove	Shi Zi Lin	狮子林
Garden of the Master of the Fishing Nets	Wang Shi Yuan	网师园
Artless Administrator's Garden	Zhuo Zheng Yuan	拙政园
(Humble Administrator's Garden,		
Garden of the Unsuccessful Politician)		
Couple's Garden	Ou Yuan	偶园
Garden to Linger In	Liu Yuan	留园
Western Garden	Xi Yuan	西园
Blue Waves Pavilion	Cang Lang Ting	沧浪亭
Garden of Pleasure	Yi Yuan	怡园
Garden of Arts	Yi Pu	艺圃
Mountain Villa Surrounded by Flourishing Greenery	Huan Xiu Shan Zhuang	环秀山庄
Crane Garden	He Yuan	鹤园
Tiger Hill	Hu Qiu	虎丘
SHANGHAI AREA		
Shanghai		上海
Garden to Please (Yu Garden)	Yu Yuan	豫园
Nanxiang		南翔
Ancient Garden of Elegance	Gu Yi Yuan	古猗园
Jiading		嘉定
Garden of Autumn Vapours	Qiu Xia Pu	秋霞圃
Nanxun		南浔
Lesser Lotus Manor	Xiao Lian Zhuang	小莲庄
Songjiang		松江
Drunken Bai Pond	Zui Bai Chi	醉白池
ZHEJIANG		浙江
Hangchow (Hangzhou)		杭州
West Lake	Xi Hu	西湖
Mansion of the Guo Family	Guo Zhuang	郭庄
(Fenyang Villa)	Fenyang Bieshu	汾阳别墅
Shaoxing		绍兴
Orchid Pavilion	Lan Ting	兰亭
Garden of the Shen family	Shen Yuan	沈园
Green Vine Studio	QingTeng Shuwu	青藤书屋
Former residence of Qiu Jin	Qiu Jin Guju	秋瑾故居
Three Flavours Studio	San Wei Shuwu	三味书屋
Ningpo		宁波
Tianyige Library	Tian Yi Ge	天一阁
ANHUI		安徽
Shexian		歙县
Xin'an Garden of Stelai	Xin'an Bei Yuan	新安碑园

WEST CHINA

SHAANXI		陕西
Xi'an		西安
Huaqing Pool	Hua Qing Chi	华清池

ENGLISH TRANSLATION	ROMANIZATION	CHARACTERS
SICHUAN		四川
Chengdu		成都
The Thatched Hut of Du Fu	Du Fu Caotang	杜甫草堂
Green Fortification Mountain	Qing Cheng Shan	青城山
SOUTH CHINA		
GUANGDONG		
Canton (Guangzhou)		广东 广州
Orchid Nursery	Lan Pu	兰圃
Panyu		番禺
Mountain Cottage of Abundant Shade	Yu Yin Shan Fang	余阴山房
Shunde		顺德
Garden of Pure Splendour	Qing Hui Yuan	清晖园
Foshan		佛山
Garden of the Liang Family	Liang Yuan	梁园
(Thatched Hall of Assembled Stars)	Qun Xing Caotang	群星草堂
Dongguan		东莞
Could-be Garden	Ke Yuan	可园
MACAO		澳门
Lou Lim-Ieok Garden	Lu Lianruo hua yuan	卢廉若花园

NOTES

Chinese names have been converted to *pinyin* romanization, with the original romanization in brackets to assist in tracing references. In the case of book and article titles, the original romanization has been retained, with the *pinyin* given in brackets.

INTRODUCTION

1. Osvald Sirén, *Gardens of China*, Ronald Press Co., New York, 1949; *China and Gardens of Europe of the Eighteenth Century*, Ronald Press Co., New York, 1950.
2. For example, Florence Ayscough, 'The Chinese Idea of a Garden', *The China Journal of Science and Arts*, vol. 1, nos. 1–4, January–July 1923; Chuin Tung [Tong Jun], 'Chinese Gardens: Especially in Kiangsu and Chekiang', *T'ien Hsia Monthly*, vol. III, no. 3, October 1936, pp. 220–244. Maggie Keswick refers to both of these.
3. On the origins of the Chinese Architectural Society, see Wilma Fairbank, *Liang and Lin: Partners in Exploring China's Architectural Past*, University of Pennsylvania Press, Philadelphia, 1994.
4. Ji Cheng, *Yuan Ye*, Zhongguo yingzao xueshe, Peking, 1933.
5. Tong Jun, *Jiangnan yuanlin zhi*, Zhongguo gongye chubanshe, Peking, 1963; 2nd edition, Zhongguo jianzhu gongye chubanshe, Peking, 1984.
6. Ji Cheng, *Yuan ye zhushi*, ed. Chen Zhi, Zhongguo jianzhu gongye chubanshe, Peking, 1981.
7. Maggie produced an illustrated history of the company: Maggie Keswick ed., *The Thistle and the Jade: A Celebration of 150 years of Jardine, Matheson & Co.*, Octopus Books Ltd., London, 1982.
8. Maggie Keswick, 'The Gardens of China', *Design Book Review*, 1985, no. 7, pp. 38–40; 'Guide to Chinese Gardens', *Architectural Design*, 56, 9 (1986), pp. 41–47.
9. L. Tjon Sie Fat & E. de Jong ed., *The Authentic Garden: A Symposium on Gardens*, The Clusius Foundation, Leiden, 1991.
10. Maggie Keswick, Judy Oberlander & Joe Wai, *In a Chinese Garden: The Art and Architecture of the Dr Sun Yat-sen Classical Chinese Garden*, The Dr Sun Yat-sen Garden Society of Vancouver, 1990.
11. For example, in a talk given at UCLA in January 1992.
12. For example, Francesca Bray, in *Technology and Gender: Fabrics of Power in Late Imperial China*, University of California Press, Berkeley, 1997, describes the Chinese garden as 'constructed not outside the house but within its walls' (p. 84).
13. The most noted study of this aspect of the garden is Rolf A. Stein, 'Jardins en miniature d'extrême Orient', *Bulletin de l'école française d'Extrême-Orient*, 1943, pp. 1–104, subsequently developed in a book, *Le Monde en petit: Jardins en miniature et habitations dans la pensée religieuse d'Extrême-Orient*, Paris, 1987, translated as *The World in Miniature: Container Gardens and Dwellings in Far Eastern Religious Thought*, translated by Phyllis Brooks, Stanford University Press, Stanford, 1990.

14. See the comments of Stanislaus Fung in 'Here and there in *Yuan ye*', *Studies in the History of Gardens & Designed Landscapes*, vol. 19, 1999, p. 38, and in 'Word and garden in Chinese essays of the Ming dynasty: Notes on matters of approach', *Interfaces* 11, 1997, p. 79.
15. Chen Congzhou, 'Shanghai de Yuyuan yu Neiyuan', *Wenwu cankao ziliao* 1957.6, pp. 34–5; Fu Xinian, 'Ji Beijing de yige huayuan', pp. 13–19; Yang Zongrong, 'Zhuozhengyuan yan'ge yu Zhuozhengyuan tuce', p. 56; Zhou Weiquan, 'Lüe tan Bishu shanzhuang he Yuanmingyuan de jianzhu yishu', pp. 8–12; Zhu Xie, 'Ji Suzhou Huiyin huayuan', pp. 32–3.
16. Perhaps the most authoritative are An Huaiqi, *Zhongguo yuanlin shi*, Tongji daxue chubanshe, Shanghai, 1991; Zhang Jiaji, *Zhongguo zaoyuan shi*, Heilongjiang renmin chubanshe, Haerbin, 1986, repr. Mingwen shuju, Taipei, 1990; and Zhou Weiquan, *Zhongguo gudian yuanlin shi*, Qinghua daxue chubanshe, Peking, 1990.
17. An Huaiqi, *Zhongguo yuanlin yishu*, Shanghai kexue jishu chubanshe, Shanghai, 1986; Jin Xuezhi, *Zhongguo yuanlin meixue*, Jiangsu wenyi chubanshe, Nanking, 1990; Yang Hongxun, 'Jiangnan gudian yuanlin yishu gailun', *Jianzhu lishi yu lilun 2*, 1981, pp. 141–161 & p. 4, 'Zhongguo gudian yuanlin yishu jiegou yuanli', *Wenwu* 1982, no. 11, pp. 49–56, *The Classical Gardens of China: History and Design Techniques*, translated by Wang Huimin, Van Nostrand Reinhold Co., New York, 1982, *Jiangnan yuanlin lun*, Shanghai renmin chubanshe, Shanghai, 1994.
18. Wang Yi, *Yuanlin yu Zhongguo wenhua*, Shanghai renmin chubanshe, Shanghai, 1990.
19. Zhang Jiaji, *Zhongguo yuanlin yishu da cidian*, Shanxi jiaoyu chubanshe, Taiyuan, 1997.
20. Ren Changtai & Meng Ya'nan, *Zhongguo yuanlin shi*, Beijing Yanshan chubanshe, Peking, 1993, repr. Wenjin chubanshe, Taipei, 1993.
21. Wei Jiazan, *Suzhou lidai yuanlin lu*, Yanshan chubanshe, Peking, 1992, repr. Wenshizhe chubanshe, Taipei, 1994.
22. Chen Zhi, *Chen Zhi zaoyuan wenji*, Zhongguo jianzhu gongye chubanshe, Peking, 1988; Chen Congzhou, *Yuanlin tancong*, Shanghai wenhua chubanshe, Shanghai, 1980, *Chun tai ji*, Huacheng chubanshe, Canton, 1985, *Lian qing ji*, Tongji daxue chubanshe, Shanghai, 1987.
23. Xie Guozhen, 'Dieshi mingjia Zhang Nanyuan fuzi shiji', *Guoli Beiping tushuguan guankan (Journal of the National Peiping Library)*, vol. 5, no. 6, pp. 12–23.
24. Cao Xun, 'Qingdai zaoyuan dieshan yishujia Zhang Ran he Beijing de

"Shanzi Zhang"', *Jianzhu lishi yu lilun* 2, 1981, pp. 116–125; 'Ji Cheng yanjiu', *Jianzhushi* (*The Architect*) 13 (1982), pp. 1–16; 'Zhang Nanyuan fuzi shi bianwu', *Zhonghua wenshi luncong*, 1985, no. 1, pp. 271–276; 'Dieshi mingjia Ge Yuliang', *Zhongguo yuanlin*, 1986, no. 2, pp. 53–54; 'Zaoyuan dashi Zhang Nanyuan – jinian Zhang Nanyuan dansheng sibai zhounian' (1), *Zhongguo yuanlin*, 1988, no. 1, pp. 21–26, (2), 1988, no. 3, pp. 2–9; see also Wu Zhaozhao, 'Ji Cheng yu Yingyuan xingzao', *Jianzhushi* (*The Architect*) 23 (1985), pp. 167–177.

25. See for example Eleanor von Erdberg, *Chinese Influence on European Garden Structures*, ed. Bremer Whidden Pond, Harvard University Press, Cambridge, Mass., 1936; Osvald Sirén, *China and Gardens of Europe of the Eighteenth Century*, Ronald Press Co., New York, 1950; Dou Wu, 'Zhongguo zaoyuan yishu zai Ouzhou de yingxiang', *Jianzhushi lunwenji*, vol. 3, 1979, pp. 104–165.

26. See Alfreda Murck & Wen Fong, *A Chinese Garden Court: The Astor Court at the Metropolitan Museum of Art*, The Metropolitan Museum of Art, New York, 1980.

27. The most notable are: Wango H.C. Weng ed., *Gardens in Chinese Art*, China House Gallery, New York, 1968; Richard M. Barnhart, *Peach Blossom Spring: Gardens and Flowers in Chinese Paintings*, The Metropolitan Museum of Art, New York, 1983; John Hay, *Kernels of Energy, Bones of Earth: The Rock in Chinese Art*, China House Gallery, New York, 1985; June Li & James Cahill, *Paintings of Zhi Garden by Zhang Hong: Revisiting a Seventeenth-Century Chinese Garden*, Los Angeles County Museum of Art, Los Angeles, 1996.

28. The issues vol.18, no. 3, Autumn 1998, and vol. 19, no. 3/4, July–December 1999.

29. Michèle Pirazzoli-t'Serstevens, Anne Chayet et al., *Le Yuanmingyuan: jeux d'eau et palais européens du XVIIIe siecle à la cour de Chine*, Éditions Recherche sur les civilisations, Paris, 1987.

30. Régine Thiriez, *Barbarian Lens: Western Photographers of the Qianlong Emperor's European Palaces*, Gordon & Breach Publishers, Amsterdam, 1998.

31. Philippe Forêt, 'The intended perception of the Imperial Gardens of Chengde in 1780', *Studies in the History of Gardens & Designed Landscapes*, vol. 19, no. 3/4, Autumn/Winter 1999, pp. 343–363; *Mapping Chengde: the Qing landscape enterprise*, University of Hawai'i Press, Honolulu, 2000.

32. Stanislaus Fung, 'Here and there in *Yuan ye*', *Studies in the History of Gardens & Designed Landscapes*, vol. 19, no. 1, Spring 1999, pp. 36–45, 'The interdisciplinary prospects of reading Yuan Ye', *Studies in the History of Gardens & Designed Landscapes*, vol. 18, no. 3, Autumn 1998, pp. 211–231; Ji Cheng, *The Craft of Gardens*, tr. Alison Hardie, Yale University Press, New Haven & London, 1988; id., *Yuanye: Le traité du jardin*, traduit du chinois et annoté par Che Bing Chiu, Les Éditions de l'Imprimeur, Besançon, 1997.

33. Ji Cheng, *Yuan ye quanshi*, ed. Zhang Jiaji, Shanxi renmin chubanshe, Taiyuan, 1993; id., *Yuanlinshuo yizhu (yuanming 'Yuan ye')*, ed. Liu Qianxian, Jilin wenshi chubanshe, Changchun, 1998; Chen Zhi's edition of the *Yuan ye* was also reprinted in Taipei by Mingwen shuju in 1983.

34. Craig Clunas, *Superfluous Things: Material Culture and Social Status in Early Modern China*, University of Illinois Press, Urbana & Chicago, 1991.

35. Philip K. Hu, 'The Shao Garden of Mi Wanzhong (1570–1628): revisiting a late Ming landscape through visual and literary sources', *Studies in the History of Gardens & Designed Landscapes*, vol. 19, no. 3/4, July–December 1999, pp. 314–342.

36. Duncan Campbell, 'Qi Biaojia's 'Footnotes to Allegory Mountain': Introduction and Translation', *Studies in the History of Gardens & Designed Landscapes*, vol. 19, no. 3/4, July–December 1999, pp. 243–275.

37. Robert E. Harrist, Jr., 'Site Names and Their Meaning in the Garden of Solitary Enjoyment', *Journal of Garden History*, 13 (1993), pp. 199–213.

38. John Makeham, 'The Confucian Role of Names in Traditional Chinese Gardens', *Studies in the History of Gardens & Designed Landscapes*, vol. 18, no. 3, Autumn 1998, pp. 187–210.

39. Georges Métailié, 'Lettrés jardiniers en Chine ancienne', *Journ. d'Agric. Trad. et de Bota. Appl.*, nouvelle série, 1995, vol. 37, no. 1, pp. 31–44; 'Some Hints on 'Scholar Gardens' and Plants in Traditional China', *Studies in the History of Gardens and Designed Landscapes*, vol. 18, no. 3, Autumn 1998, pp. 248–256; Georges Métailié & Nicole Staüble Tercier, *Au jardin potager chinois*, Alimentarium Vevey, 1997.

40. Peter Valder, *The Garden Plants of China*, Weidenfeld & Nicolson, London, 1999; *Gardens in China*, Timber Press, Portland, Oregon, 2002,

41. Xu Yinong, *The Chinese city in space and time: the development of urban form in Suzhou*, University of Hawai'i Press, Honolulu, 2000.

42. Craig Clunas, *Fruitful Sites: Garden Culture in Ming Dynasty China*, Reaktion Books, London, 1996.

43. There are exceptions, such as Liu Tuo, 'Songdai yuanlin de yishu fengge', *Meishu shilun* 4, 1986, pp. 39–49, and Wang Juyuan, 'Suzhou Ming Qing zhaiyuan fengge de fenxi', *Yuanyi xuebao* (*Acta Horticulturalia*), vol. 11, no. 2, May 1963, pp. 177–194. However, even such studies as these tend

44. Antoine Gournay, 'The Yuyin Shanfang in Panyu: A Case Study', unpublished paper presented at the New England Conference of the Association for Asian Studies, 9 October 1999, discusses the European influence on this garden, one of the 'four famous gardens of Guangdong'.

45. James Turner, *The Politics of Landscape: Rural Scenery and Society in English Poetry 1630-1660*, Oxford University Press, Oxford, 1979; John Barrell, *The Dark Side of the Landscape: the Rural Poor in English Painting 1730–1840*, Cambridge University Press, Cambridge, 1980; Denis E. Cosgrove, *Social Formation and Symbolic Landscape*, Croom Helm, London & Sydney, 1984; see also Ann Bermingham, *Landscape and Ideology: The English Rustic Tradition, 1740–1860*, Thames & Hudson, London, 1986, and Tom Williamson, *Polite Landscapes: Gardens and Society in Eighteenth-Century England*, Sutton Publishing Ltd., Stroud, 1995.

46. Joanna F. Handlin Smith, 'Gardens in Ch'i Piao-chia's Social World: Wealth and Values in Late-Ming Kiangnan', *Journal of Asian Studies*, vol. 51, no. 1, February 1992, pp. 58–81.

47. John W. Dardess, 'A Ming Landscape: Settlement, Land Use, Labor, and Estheticism in T'ai-ho County, Kiangsi', *The Harvard Journal of Asiatic Studies*, vol. 49, no. 2, December 1989, pp. 295–364; id., *A Ming Society: T'ai-ho County, Kiangsi, fourteenth to seventeenth centuries*, University of California Press, Berkeley, Los Angeles & London, 1996.

48. See note 42.

49. Richard Vinograd, 'Family Properties: Personal Context and Cultural Pattern in Wang Meng's *Pien Mountains* of 1366', *Ars Orientalis*, vol. 13, 1982, pp. 1–29; Robert E. Harrist, Jr., 'Art and Identity in the Northern Sung Dynasty: Evidence from Gardens', in Maxwell K. Hearn & Judith G. Smith ed., *Arts of the Sung and Yüan*, The Metropolitan Museum of Art, New York, 1996, pp. 147–164; id., *Painting and Private Life in Eleventh-Century China: Mountain Villa by Li Gonglin*, Princeton University Press, Princeton, 1998.

WESTERN REACTIONS

1. Taken from 'A particular account of the Emperor of China's gardens near Peking in a letter from Père Attiret . . . to his friend at Paris, translated from the French by Sir Harry Beaumont' (1749).

2. Ibid., pp. 9–10.

3. Ibid., pp. 16–17.

4. Ibid., p. 5.

5. Ibid., p. 38. The translator adds a footnote here on some engravings of the Emperor's gardens he had recently seen. He notes that, although the disposition of ground, water and plantations was 'indeed quite regular', the architecture was all of a regular kind. So unlike his expectations were these drawings that he concluded Père Attiret must have been describing a different garden. He was wrong. The engravings would have been those taken from the 'Forty Views' of the *Yuan Ming Yuan*, commissioned by the Qianlong Emperor in 1749, and later sent to Paris where they can still be seen in the Bibliothèque Nationale. These scenes were exactly those described by Attiret, which only goes to show how misleading verbal descriptions can be.

6. In fact it is remarkable that Father Attiret managed to get any coherent picture of the gardens at all, although the Jesuits were allowed out of their studio only when whatever they had to paint was too large to be brought to them. They were put under a large guard of eunuchs and obliged 'to go quick, and without any noise, and huddle and steal along softly, as if . . . on some Piece of Mischief'. Even so they saw a lot more of the gardens than most of the Emperor's own subjects.

7. Hunt and Willis, *The Genius of the Place, p.* 99. See also p. 142 for the Spectator no. 414, (25 June 1712), in which Joseph Addison also took up this idea.

8. In the 'Epistle to Lord Burlington' (1731), Hunt and Willis, op. cit., p. 212.

9. Stephen Switzer, 'Ichonographica Rustica' (1718 and 1742), Hunt and Willis, op. cit., p. 152.

10. Sirén, *Gardens of China, p.* 4.

11. Chuin Tung [Tong Jun] in *T'ien Hsia Monthly* III, 3 (October 1936), p. 222.

12. Oliver Goldsmith, *The Citizen of the World,* or 'Letters from a Chinese philosopher residing in London to his friends in the country', vol. 1, letter XIV, Crookes (ed.) (1799), p. 46.

ORIGINS OF GARDENS

1. For a discussion of Shang Di [Shang Ti] and the cosmic egg, see Wolfram Eberhard, *A History of China* (London, 1960) p. 23. Michael Sullivan also mentions the idea in his *Short History of Chinese Art* (London, 1967).

2. David Hawkes (trans.) 'The Songs of Ch'u' (Ch'u Tz'u) [Chu Ci] in Cyril Birch (ed.), *Anthology of Chinese Literature* (New York, 1965) p. 75.

3. Ibid., p. 76.

4. E.R. Hughes (trans.), *Two Chinese Poets, vignettes of Han life and thought* (Princeton, 1960); Ban Gu [Pan Ku] on the Western Capital, p. 33; Zhang Heng [Chang Heng] on the Western Capital, p. 42.
5. From Sima Xiangru [Ssu-ma Hsiangju], 'Shang-lin fu' [Shanglin fu], translated by Burton Watson in Birch (ed.), *Anthology of Chinese Literature, p.* 148.
6. Hughes, op. cit., p. 73.
7. Ibid., p. 29.
8. Sima Xiangru [Ssu-ma Hsiang-ju], 'Shang-lin fu' [Shanglin fu], in Birch (ed.), op. cit., p. 145.
9. David Hawkes (trans.) *Ch'u Tz'u [Chu Ci], Songs of the South* (Oxford, 1959).
10. Rolf Stein, 'Jardins en miniature d'extrême-Orient', *Bulletin d'Ecole Française d'Extrême-Orient* (1943) p. 42.
11. Hughes (trans.), op. cit., Zhang Heng [Chang Heng] on the Western Capital, p. 39.
12. Hughes, ibid., Ban Gu [Pan Ku] on the Western Capital, p. 33.
13. Hughes, ibid., Ban Gu [Pan Ku] on the Western Capital, p. 33.
14. For a discussion of the rustic ideal see Lin Yu-tang, *My Country and My People,* pp. 34–7 and 113–14.
15. See Edward H. Schafer, *Tu Wan's Stone Catalogue* (Berkeley, 1961) p. 5, and Lorraine Kuck, *The World of the Japanese Garden* (New York, 1968) p. 95.

IMPERIAL GARDENS

1. From *P'u T'ung Hsin Li Shih [Putong xin lishi]*, a textbook on Chinese history, p. 57, quoted from Malone's *History of the Peking Summer Palaces* (Urbana, 1934) p. 192.
2. D. C. Lau (trans.), *Mencius* (London, 1970) p. 113.
3. Emperor Huizong of Song, justifying the building of his garden in the *Record of Gen Yue* (1122) trans. from the Chinese by Grace Wan. In *The Vermilion Bird* Schafer explains the word *ling* as 'a spiritual force or energy emanating from or potential in any object full of mana, even such a dull thing as a stone'; he notes that it 'radiated abundantly from Imperial tombs, from powerful drugs, from magical animals and from haunted trees'. For King Wen's park see Arthur Waley's translation of the *Shi Jing: The Book of Songs* (London, 1969) no. 274, p. 259.
4. Both this and subsequent quotations are from 'Sui Yangdi hai shan ji' (Sea and Mountain Records of Sui Yangdi) in the *Tang Song chuanqi ji* (collection of fictional works of the Tang and Song dynasties). Reprinted in *Zhongguo yingzao xue* (Bulletin of the Society for Research in Chinese Architecture) vol. IV, June 1934. Trans. from the Chinese by Alexander Soper in Lorraine Kuck, *The World of the Japanese Garden* (New York, 1968) p. 19.
5. The source of this quote has unfortunately been lost in the mail somewhere between Los Angeles and London.
6. See the *Sui shu* (Annals of the Sui). For a discussion of Sui Yangdi see Arthur F. Wright, 'Sui Yang-ti [Sui Yangdi], Personality and Stereotype', in *The Confucian Persuasion* (Stanford, 1960).
7. Murakami Yoshimura, 'Imperial Gardens of Ch'ang-an in the T'ang dynasty', *History* no.3 (Kansai University, 1955), trans. from the Japanese by Keiko Ryare.
8. Joseph Needham, *Science and Civilization in China* (Cambridge, England, 1954 (p. 65), vol. 4 (1965), part 2, p. 134. This chapter also discusses other imperial waterworks, and the 'hydraulic elegances' *(shui shi)* of Sui Yangdi, p. 160. Plate CLXVII shows an artificial channel, at Kyongju in Korea, for floating cups at poetic garden parties. Sirén has a photograph of a similar wine-cup stream in *Gardens of China*, plate 22.
9. Edward Schafer, *The Golden Peaches of Samarkand* (Berkeley and Los Angeles, 1967) p. 233. See also pp. 212 and 179 for notes on Ming Huang's great nursery and herb gardens; p. 120 for an account of a game of go (Chinese chess) in the garden; and p. 92 for a note on the trouble caused by eunuchs sent by Ming Huang to collect duck south of the Yangtze for the imperial park in Chang'an.
10. Both quotes are from the *Record of Hua Yang Palace* by the monk Zixiu, private translation by Grace Wan.
11. Ibid.
12. Both quotes are from the *Record of Gen Yue* by Emperor Huizong, trans. by Grace Wan.
13. ixiu, op. cit.
14. This and subsequent quotes are from Marco Polo, *The Travels,* trans. by Ronald Latham, new edn. (London, 1972) p. 225.
15. Ibid., p. 127.
16. Carroll Brown Malone, *History of the Peking Summer Palaces* (Urbana, 1934) p. 64. Translation from the introduction to *Yu zhi Yuan Ming Yuan sishi jing shi* (Imperial Poems on the Forty Scenes of the Yuan Ming Yuan).
17. Ibid., p. 113.
18. George Kates, *The Years That Were Fat* (Cambridge, Mass., 1967) p. 210.
19. Marquis Zeng [Tseng] in *Asiatic Monthly Review,* quoted by Malone, *History of the Peking Summer Palaces* op. cit., p. 192.
20. Su Hua, quoting the old family gardener Lao Zhou [Chou], in *Ancient Melodies,* (London, 1969), chapter 12, p. 169.

THE GARDENS OF THE LITERATI

1. See Fung Yu-lan, *A Short History of Chinese Philosophy,* trans. by D. Bodde (London, 1948) p. 235.
2. The exact source of this quotation has unfortunately been lost in the mail somewhere between Los Angeles and London. However for a similar idea see the *Zhuangzi [Chuang Tzu]*, trans. by Legge, *The Texts of Taoism* (Oxford, 1891) vol. 2, p. 59.
3. Herlee G. Creel, *What is Taoism?* (Chicago, 1970), p. 2.
4. *Zhuangzi [Chuang Tzu]*, trans. Legge, *The Texts of Taoism* (Oxford, 1891), chapter 22, vol. 2, p. 59.
5. From A. C. Graham's translation of *The Book of Lieh Tzu [Liezi]*, (London, 1960) p. 44. See also another version of the story in the *Zhuangzi [Chuang Tzu]*, trans. Herbert A. Giles (London, 1964) chapter 19, p. 184.
6. This is Joseph Needham's interpretation in *Science and Civilization in China*, vol. 2, p. 68. For a long essay on the origins and meanings of *wu wei*, see Creel, *What is Taoism?* op. cit.
7. Creel, *What is Taoism?* op. cit.
8. A favourite image of the *Dao De Jing;* stone connoisseurs of the Tang and later dynasties may also have regarded their rocks, strangely shaped by natural forces, as symbols of the 'uncarved block'.
9. Trans. by Wolfram Eberhard in *A History of China* (London, 1960) p. 49.
10. See J. D. Frodsham, *The Murmuring Stream* (Kuala Lumpur, 1967) p.1.
11. Ibid., p. 2. And see also his 'Origins of Chinese Landscape Poetry', *Asia Major* VIII part I (1960) pp. 68–103.
12. From Xie Lingyun's *Fu on Dwelling in the Mountains* in Frodsham, *The Murmuring Stream,* p. 43.
13. Tao Qian [T'ao Ch'ien], 'Returning to the Fields', trans. Arthur Waley, *Chinese Poems,* (London, 1946) p. 92.
14. Tao Qian [Tao Ch'ien], preface (dated the 11th moon of the year 405) trans. by J. R. Hightower, *The Poetry of Tao Ch'ien [Tao Qian]* (Oxford, 1970) no. 62, p. 268.
15. Ibid., p. 269.
16. Arthur Waley *(trans.),Chinese Poems,* pp. 91–92.
17. Arthur Waley, *The Life and Times of Po Chu-i* [Bai Juyi] (London, 1949) and see also Sirén, *Gardens of China, p.* 75.
18. 'The Recluse Pavilion', trans. Lin Yutang, *The Gay Genius* (1948) p. 154.
19. Sima Guang 'Record of the Park of Solitary Enjoyment' from *Sima Wen gong wen ji* (Taipei, 1967), translated for me by Paul Clifford. See also Sirén, *Gardens of China, p.* 77 for another translation.
20. Shen Fu, *Chapters from a Floating Life,* trans. Shirley M. Black (London, 1960).
21. Chen Haozi [Ch'en Hao-tzu], *The Mirror of Flowers* (1688), trans. by Lin Yu-tang.
22. Derek Bodde in his introduction to Bradley Smith and Wang-go Weng, *China – A History in Art* (London, 1973).
23. Bai Juyi [Po Chü-I], 'The Grand Houses at Loyang' [Luoyang], written c. AD 829. Trans. Arthur Waley, *Chinese Poems,* (London, 1971) p. 162.
24. Li Gefei, *Luoyang mingyuan ji [The Celebrated Gardens of Luoyang]*, translated for me by Lillian Chin.
25. Ibid.
26. For example, in the sixteenth-century erotic novel, *Jin Ping Mei,* translated by Clement Egerton as *The Golden Lotus,* 4 vols., (London, 1972), the 'hero' takes a party of friends on an outing to view the famous garden of an immensely rich eunuch. When Shen Fu and his wife visit the Cang Lang Ting in *Six Chapters From a Floating Life,* they have the keeper of the gate prevent anyone else from coming in for the evening. In the *Yuan Ye,* Ji Cheng warns his readers about the improprieties of writing spontaneous verse on other people's walls while visiting gardens.

THE PAINTER'S EYE

1. *Dao De Jing,* Book 1, no. XII. Later the Neo-Confucianists worked out a singularly Chinese attitude to the enjoyment of sensual objects: as long as the sage merely *rested* and did not fix his mind on them, sensual objects would be delightful but not obsessive or corrupting. Later scholars aimed to appreciate, for instance, a beautiful painting as an imprint of the maker's soul, without becoming attached to the object *per se.*
2. Wan-go Weng, *Gardens in Chinese An* (New York, 1968).
3. L. Sickman and A. Soper, *The Art and Architecture of China* (Harmondsworth, 1971), p. 133 and note p. 480. For another full discussion of Xie He's [Hsieh Ho] six principles see Sirén, *The Chinese on the Art of Painting,* pp. 18–22.
4. To see why this vagueness and ambiguity were so fruitful we only have

to analyse the phrase's separate parts. *Qi* means roughly 'life breath', 'spirit', 'vitality', 'activity of the spirit'. It is not, says Sirén, 'an abstract concept, but an actual phenomenon', and it 'animates everything in nature – including human beings'; *yun* means 'resonance', 'consonance', 'harmonized vibrations'; *sheng* means 'life movement'; while *dong* means 'physical motion' or 'animation'. Thus the phrase could be rendered as 'vibration of the vitalizing spirit and movement of life'. One is reminded of vitalist philosophies in this century, *élan vital* etc., not to mention the hippy emphasis on judging things according to their 'good vibes'. In the seventeenth-century garden manual, the classic *Yuan Ye* of the Ming dynasty, the phrase means according to Osvald Sirén, something like 'the breath of Nature's own pulsating life'. Hence magic, psychology, realism and metaphysics are all mixed up and summarized in this key principle. Also see Michael Sullivan, *The Birth of Landscape Painting in China, p. 106,* for another discussion, listing alternative translations.

5. For a discussion of Zhang Yanyuan's [Chang Yen-yuan] *Lidai minghua ji,* see Sirén, *The Chinese on the Art of Painting,* pp. 18–28 and 30.
6. See Sickman and Soper, *The Art and Architecture of China,* pp. 203–4.
7. Chuin Tung [Tong Jun], *T'ien Hsia Monthly,* III, 3 (October 1936), pp. 222–3.
8. For Guo Xi's [Kuo Hsi] *Essay on Landscape Painting,* see trans. by Shio Sakanishi, first pub. London, 1935; reprinted 1939; also Sirén, *Chinese Painting* (New York, 1963), vol. I, pp. 215–30.
9. Sickman and Soper, *The Art and Architecture of China, p. 219,* or Sakanishi, op. cit., p. 41.
10. Sirén, *Gardens of China.*
11. See James Cahill, *Chinese Painting* (Geneva, 1960), pp. 62–4.
12. For an alternative translation see Susan Bush, *The Chinese Literati on Painting* (Cambridge, Mass., 1971), p. 31.
13. Quoted in James Cahill, *Chinese Painting* (Geneva, 1960), p. 113.
14. Sirén, *Gardens of China,* frontispiece, plates 92–5, and discussion pp. 80–1.
15. Li Yu, *Li Yu quan ji* (Taipei, 1970), section V.
16. Mo Shilong, *Hua shuo (Baoyantang biji),* reprint (Shanghai, 1922). See Sirén, *Chinese Painting, vol. 5,* p. 10 et seq.
17. In a private conversation in New York.
18. Sirén, *The Chinese on the Art of Painting, p. 183.*

ARCHITECTURE IN GARDENS

1. Trans. by Lin Yu-tang in *My Country and My People, p. 316.*
2. Tong Jun, *Jiangnan yuanlin zhi* (Record of Gardens South of the Yangtze), chapter 1, translated for me by Kenneth Ma.
3. This and the following quotation are translated and paraphrased by Florence Ayscough in 'The Chinese idea of a garden' from *A Chinese Mirror* (London, 1925) pp. 19, 29. See also a full account of the Yu Yuan garden in Chinese by Chen Congzhou, *Wen Wu* (Peking, 1957) no. 6, pp. 34, 35.
4. Tong Jun, op. cit., is paraphrasing the aesthetics of the eighteenth-century writer Shen Fu. (See trans. in Lin Yutang, *My Country and My People,* p. 312.
5. In his 'Informal Records of Random Thoughts' [Xianqing ouji] (see *Li Yu quan ji,* Taipei, 1970) Li Yu also claims to have invented the *trompe l'oeil* window filled in with real, but dead branches. Their advantage is that they may mask an unsatisfactory view, and they can be changed from time to time. They were also cheap to make as the impoverished connoisseur could arrange them himself.
6. Trans. Sirén, *Gardens of China.*
7. Tong Jun, op. cit.
8. Translated for me by Paul Clifford. Yuan Mei believed the Sui Yuan to have been the model for the garden in the *Dream of the Red Chamber.* For a full account of his life see Arthur Waley's *Yuan Mei, 18th-Century Chinese Poet* (London, 1956).
9. Attiret's letter to M. d'Assaut. The translation here is from Hope Danby, *The Garden of Perfect Brightness* (London, 1950) pp. 69–78.
10. After the *Yuan Ye* (1631), the only classification I have found of Chinese lattice work is the well-known work of Daniel Sheets Dye, *A Grammar of Chinese Lattice* (Cambridge, Mass., 1937). Neither book tries to be comprehensive, and neither shows designs that are exclusive to gardens, but the Sheets Dye in particular gives a good sense of their remarkable complexity and richness.
11. This and subsequent quotes are from David Hawkes' translation, *The Story of the Stone* (Harmondsworth, 1973) pp. 324–347.

ROCKS AND WATER

1. Description by Yuan Xuanzhi [Yuan Hsüan-chih], trans. by Alexander Soper in the end-notes to Michael Sullivan, *The Birth of Landscape Painting in China, p. 197.* Zhang Lun was Minister of Agriculture under

the Emperor Xiaomingdi (AD 516–27) of the Northern Wei dynasty, and built his garden in Luoyang.
2. Li Yu, 'Informal Records of Random Thoughts' [Xianqing ouji], in *Li Yu quan ji* (Taipei, 1970) section 5, part 1, translated for me by P. D. Lu.
3. Osvald Sirén, *Gardens of China, p. 26.*
4. Edward H. Schafer, *Tu Wan's* [Du Wan] *Stone Catalogue* (Berkeley, 1961) pp. 52–53.
5. Arthur Waley, *Analects of Confucius* (London, 1938) Book VI, number 21, p. 121.
6. Yang Hongxun, *Wen Wu* (Peking, 1957) number 6.
7. Ibid.
8. *Dao De Jing,* Book I, VIII. Translated by Joseph Needham, *Science and Civilization in China* (Cambridge, England, 1954–65), vol. 2 (1956), p. 57.
9. See note on chapter 3, no. 5.
10. *Zhuangzi* [Chuang Tzu], trans. H. A. Giles (London, 1964) chapter V, p. 68.
11. Rolf Stein, 'Jardins en miniature d'extrême-Orient', *Bulletin d'Ecole Française d'Extrême-Orient,* 1943.
12. Edward H. Schafer, *The Divine Woman, p. 7.*
13. Ibid., p. 29.
14. Marco Polo, *The Travels of Marco Polo,* trans. by Ronald Latham (London, 1972) p. 219.
15. See Arthur Waley, *The Life and Times of Po Chü-i* [Bai Juyi] (London, 1949) p. 149.

FLOWERS, TREES AND HERBS

1. This and the subsequent quotations are from Ernest Wilson, *China, Mother of Gardens, p. 37.* See also Chapter XXV 'Gardens and Gardening', pp. 322–328 for a discussion on plants in Chinese gardens. For more on plant collecting in China see Robert Fortune's accounts of his travels, *A Journey to the Tea Countries* (London, 1852, repr. 1987), and E.H.M. Cox's history of botanical exploration: *Plant Hunting in China* (London, 1945).
2. See note on Ch. 3, no. 5.
3. Zhou Dunyi 'Ai lian shuo' [On the love of lotuses], *Zhou Lianxi ji,* 8, 139.
4. Edward Schafer, *The Golden Peaches of Samarkand* (Berkeley and Los Angeles, 1967), p. 119. Schafer also makes an interesting parallel with the use of homely plants in English poetry: 'Geoffrey Grigson has shown how native plants are used in English poetry to reflect and stimulate the deeper human emotions, while exotics, lacking the long and intimate interrelation with the English people, can do little more for a poet than make an exciting and colourful splash in his verses. So it was in China'.
5. Trans. by Arthur Waley, *Book of Songs,* (London, first ed., 1937), p. 35.
6. Trans. by Robert Kotewell and Norman L. Smith in *The Penguin Book of Chinese Verse* (Harmondsworth, England, 1962), p. 68.
7. In *The Vermilion Bird, p. 195,* Schafer notes that *gui* [*kuei*] was used for *cinnamomum cassia* and other members of the laurel family, and that 'poets took note of the *kuei* chiefly for its rich aroma'. A cassia tree was anciently thought to grow on the moon, 'so it was natural for a sophisticated man of medieval China to admire any *kuei* by moonlight'. It is this moon-*gui* that Schafer identified as the delicately scented osmanthus. He also notes that the great official Li Deyu brought an exotic red *gui* back with him from exile in the south, to plant in his famous Ping Quan villa near Luoyang.
8. Local tisanes can still be bought at most beauty-spots in modem China, including chrysanthemum tea, and a delicious pink brew which is a speciality of the Seven-star Crags in Guangdong province.
9. Schafer in *The Golden Peaches of Samarkand, p. 128,* quotes Bai Juyi's [Po Chü-i] poem on Lotus-gathering ladies from *Bai shi chang qing ji,* 28, 76.
10. Shen Fu, *Chapters From a Floating Life,* trans. Shirley M. Black, p. 71. The old Qing Empress Cixi also liked lotus-scented tea produced this same way, see Su Hua, *Ancient Melodies* (London, 1969). For the economic aspects of elaborate gardens see the *Dream of the Red Chamber* where, as the family fortunes begin to slide, the produce of the Da Guan Yuan becomes a source of revenue (ch. 56).
11. An English translation of this poem can be found by John A. Turner in *A Golden Treasury of Chinese Poetry* (Hong Kong, 1976), p. 111, and by Shigeyoshi Obata in *The Works of Li Po* [Li Bai] (New York, 1922), p. 31.
12. 'The Flower Market', trans. Arthur Waley in *Chinese Poems* (London, 1971), p. 121.
13. 'Chance verse on a summer day', by Huang Youzao [Huang Yu-tsao], from Xie Wuliang, *Zhongguo funü wenxue shi* [A History of Chinese Women's Literature], ch. 8, p. 49. English translation by Robert Kotewall and Norman L. Smith in *The Penguin Book of Chinese Verse* (Harmondsworth, 1971) p. 61.
14. Lorraine Kuck, *The World of the Japanese Garden* (New York, 1968), p. 33.
15. Shen Fu, *Chapters From a Floating Life,* trans. Black, p. 33.
16. Sitwell, 'Old Worlds for New', from *Queen Mary and Others* (London, 1974).
17. Shen Fu, *Chapters From a Floating Life,* trans. Black, pp. 59–60.

18. Su Hua, *Ancient Melodies* (London, 1969), pp. 173–174.

MEANINGS OF THE CHINESE GARDEN

1. See appendix for gardens that can be visited in mainland China. In addition there are a smattering of Chinese-inspired gardens elsewhere: in Napa Valley, California, there is a simple but effective lake garden at the Montelina Winery with willows, an island, bridge and pavilion. Views of the surrounding hills are 'borrowed', in the manner of the West Lake, Hangzhou. In Staffordshire, England, there is a Chinoiserie 'Chinese garden' at Biddulph Grange, part of a remarkable complex that includes also 'Japanese' and 'Egyptian' gardens. Some other eighteenth-century English gardens have so called 'Chinese' pavilions. Like the one at Shugborough, Staffordshire or the Woburn Dairy they may be charming but they have nothing the slightest Chinese about them. In the Metropolitan Museum of Modern Art in New York there is the Astor Court, in the centre of the galleries of Chinese art, meticulously built and erected with great attention to authenticity in co-operation with the Chinese. It is a sparse, cool space in the heart of New York City, but lacks the intensity and rigour of a real Chinese garden. In Maine, USA, there is the Abby Aldrich Rockefeller garden, designed by Beatrix Farrand around a collection of stone Korean lanterns and tomb figures. A pink wall in Chinese style encloses an outer garden where the sculptures are set informally among trees and mosses in a somewhat Japanese style. An inner wall entered through a vase door leads into an open garden with English-inspired borders round a large rectangular lawn. A moon gate in the wall at the far end focuses the scene on an old pine tree. The garden is an inspired piece of landscape design but though it borrows Chinese motifs, it neither intends, nor achieves, any feeling of a real Chinese Garden.

In Macao there is an interesting tropical Chinese garden, with much Portugese influence, called the Lou Lim Ieoc Garden. Renovated in 1972, it is now a public park. There are also plans, in Taiwan to rebuild Ban Qiao, garden of the magnificent Lin family compound in the suburbs of Taipei. It is now [1978] in ruins, however, and getting yearly worse while the project remains only a good intention. Two interesting new Chinese gardens, built with help from China, are now being laid out in Vancouver and San Francisco. A third is proposed for Houston, Texas.

2. Quotes are from David Hawkes' translation of the *Dream of the Red Chamber*, (*The Story of the Stone*), pp. 334–7.

3. Andrew H. Plaks, *Archetype and Allegory*, p. 186. On the previous page he gives a somewhat different translation to Hawkes, of the same dispute.

4. Ibid., p. 146.

5. Ibid., p. 50

6. The translation is by Plaks, op. cit., p. 167.

7. Ibid., p. 168.

8. For instance Lawrence Sickman, in writing about landscape painting, says '... the Chinese spirit eventually found in the world of nature an answer to its longings that was more real and gratifying than the pomp of Imperial courts, Confucian morality, or all the multiple gods of the Mahayana Buddhist pantheon'. See Sickman and Soper, *Art and Architecture of China*, p. 186. See also Sherman Lee, *Chinese Landscape Painting*, p. 3, for a discussion of the importance of nature philosophy.

9. See Edmund Leach, *Culture and Communication*, 'The logic by which symbols are connected', (Cambridge, 1976), pp. 33–6, 72, 82–3. This short 'structuralist guidebook' is very suggestive in its analysis and speculation on various forms of myth, ritual and social behaviour.

10. Nelson I. Wu, *Chinese and Indian Architecture*, pp. 45–6.

BIBLIOGRAPHY

listing books and articles on Chinese gardens, or including discussions of the subject, published in English and other European languages since the first publication of *The Chinese Garden*, together with some earlier books and articles not referred to by Maggie Keswick. Books in Chinese are not included, but some details have been given in the footnotes. Readers of Chinese are also referred to Stanislaus Fung's 'Guide to secondary sources on Chinese gardens'.

Richard M. Barnhart, *Peach Blossom Spring: Gardens and Flowers in Chinese Paintings*, The Metropolitan Museum of Art, New York 1983.

Marianne Beuchert, *Die Gärten Chinas*, Diederichs, Köln, 1983, 2nd edition: Munich, 1988.

Francesca Bray, *Technology and Gender: Fabrics of Power in Late Imperial China*, University of California Press, Berkeley, 1997.

Duncan Campbell, 'Qi Biaojia's 'Footnotes to Allegory Mountain': Introduction and Translation', *Studies in the History of Gardens & Designed Landscapes*, vol. 19, no. 3/4, July–December 1999, pp. 243–275.

Chen Congzhou, *Shuo yuan/On Chinese Gardens*, translated by Chen Xiongshan et al., Tongji University Press, Shanghai, 1984.

Chen Haozi, *Miroir des fleurs: guide pratique de jardinier amateur en Chine au XVIIe siecle*, translated by Jules Halphen, Plon, Paris, 1900.

Chen Lixian, *Art and Architecture in Suzhou Gardens*, Yilin Press, Nanking, 1992.

Chiu Che Bing, 'Yuanye, le traité du jardin chinois', *Asies II – Aménager l'espace*, ed. Flora Blanchon, Presses de l'Université de Paris-Sorbonne, Paris, 1993, pp. 281–296.

— 'Droiture et Clarté: scène paysagère au Jardin de la Clarté Parfaite', *Studies in the History of Gardens and Designed Landscapes*, vol. 19, no. 3/4, Autumn/Winter 1999, pp. 364–375.

Chiu Che Bing & Gilles Baud Berthier, *Yuanming Yuan: le jardin de la clarté parfaite*, Les Éditions de l'Imprimeur, Besançon, 2000.

Chung Wah Nan, *The art of Chinese gardens*, Hong Kong University Press, Hong Kong, 1982.

Craig Clunas, *Superfluous Things: Material Culture and Social Status in Early Modern China*, University of Illinois Press, Urbana & Chicago, 1991.

— *Fruitful Sites: Garden Culture in Ming Dynasty China*, Reaktion Books, London, 1996.

— 'Ideal and reality in the Ming garden', *The Authentic Garden: A Symposium on Gardens*, ed. L. Tjon Sie Fat & E. de Jong, The Clusius Foundation, Leiden, 1991, pp. 197–205.

— 'The Gift and the Garden', *Orientations*, vol. 26, no. 2, February 1995, pp. 38–45.

J.C. Cooper, 'The Symbolism of the Taoist Garden', *Studies in Comparative Religion*, Autumn 1977, pp. 224–234.

John W. Dardess, 'A Ming Landscape: Settlement, Land Use, Labor, and Estheticism in T'ai-ho County, Kiangsi', *The Harvard Journal of Asiatic Studies*, vol. 49, no. 2, December 1989, pp. 295–364.

— *A Ming Society: T'ai-ho County, Kiangsi, fourteenth to seventeenth centuries*, University of California Press, Berkeley, Los Angeles & London, 1996.

Eleanor von Erdberg, *Chinese Influence on European Garden Structures*, ed. Bremer Whidden Pond, Harvard University Press, Cambridge, Mass., 1936.

Wilma Fairbank, *Liang and Lin: Partners in Exploring China's Architectural Past*, University of Pennsylvania Press, Philadelphia, 1994.

Philippe Forêt, *Mapping Chengde: the Qing landscape enterprise*, University of Hawai'i Press, Honolulu, 2000.

— 'The intended perception of the Imperial Gardens of Chengde in 1780', *Studies in the History of Gardens & Designed Landscapes*, vol. 19, no. 3/4, Autumn/Winter 1999, pp. 343–363.

Stanislaus Fung, 'Word and garden in Chinese essays of the Ming dynasty: Notes on matters of approach', *Interfaces* 11, 1997, pp. 77–90.

— 'Notes on the Make-do Garden', *Utopian Studies*, vol. 9, no. 1, 1998, pp. 142–148.

— 'The interdisciplinary prospects of reading Yuan Ye', *Studies in the History of Gardens & Designed Landscapes*, vol. 18, no. 3, Autumn 1998, pp. 211–231.

— 'Guide to secondary sources on Chinese gardens', *Studies in the History of Gardens & Designed Landscapes*, vol. 18, no. 3, Autumn 1998, pp. 269–286.

— 'Here and there in *Yuan ye*', *Studies in the History of Gardens & Designed Landscapes*, vol. 19, 1999, pp. 36–45.

— 'The Imaginary Garden of Liu Shilong', *Terra Nova*, vol. 2, no. 4, pp. 15–21.

Antoine Gournay, 'L'aménagement de l'espace dans le jardin chinois', *Asies II – Aménager l'espace*, ed. Flora Blanchon, Presses de l'Université de Paris-Sorbonne, Paris, 1993.

David L. Hall & Roger T. Ames, 'The Cosmological Setting of Chinese Gardens', *Studies in the History of Gardens & Designed Landscapes*, vol. 18, no. 3, Autumn 1998, pp. 175–186.

Kenneth James Hammond, 'Wang Shizhen's Yan Shan Garden essays: narrating a literati landscape', *Studies in the History of Gardens & Designed Landscapes*, vol. 19, no. 3/4, July–December 1999, pp. 276–287.

Joanna F. Handlin Smith, 'Gardens in Ch'i Piao-chia's Social World: Wealth and Values in Late-Ming Kiangnan', *Journal of Asian Studies*, vol. 51, no. 1, February 1992, pp. 58–81.

Alison Hardie, 'Ji Cheng's *Yuan Ye* (*The Craft of Gardens*) in its social setting', *The Authentic Garden: A Symposium on Gardens*, ed. L. Tjon Sie Fat & E. de Jong, The Clusius Foundation, Leiden, 1991, pp. 207–214.

— 'Hu Yinglin's 'Connoisseurs of Flowers': translation and commentary', *Studies in the History of Gardens & Designed Landscapes*, vol. 19, no. 3/4, Autumn/Winter 1999, pp. 272–275.

James M. Hargett, 'Huizong's Magic Marchmount: the Genyue Pleasure Park of Kaifeng', *Monumenta Serica*, 38 (1989–1990), pp. 1–48.

Robert E. Harrist, Jr., *Painting and Private Life in Eleventh-Century China: Mountain Villa by Li Gonglin*, Princeton University Press, Princeton, 1998.

— 'Art and Identity in the Northern Sung Dynasty: Evidence from Gardens', in Maxwell K. Hearn & Judith G. Smith ed., *Arts of the Sung and Yüan*, The Metropolitan Museum of Art, New York, 1996, pp. 147–164.

— 'Site Names and Their Meaning in the Garden of Solitary Enjoyment', *Journal of Garden History*, 13 (1993), pp. 199–213.

John Hay, *Kernels of Energy, Bones of Earth: The Rock in Chinese Art*, China House Gallery, New York, 1985.

Hu Dongchu, *The Way of the Virtuous: The Influence of Art and Philosophy on Chinese Garden Design*, New World Press, Peking, 1991.

Philip K. Hu, 'The Shao Garden of Mi Wanzhong (1570-1628): revisiting a late Ming landscape through visual and literary sources', *Studies in the History of Gardens & Designed Landscapes*, vol. 19, no. 3/4, July–December 1999, pp. 314–342.

William Hung Yeh, *The Mi Garden (Shaoyuan tulu kao)*, Harvard-Yenching Institute Sinological Index Series, Supplement no. 5, Peking, 1933.

Ji Cheng, *The Craft of Gardens*, tr. Alison Hardie, with a foreword by Maggie Keswick, Yale University Press, New Haven & London, 1988.

— *Yuanye: Le traité du jardin*, traduit du chinois et annoté par Che Bing Chiu, Les Éditions de l'Imprimeur, Besançon, 1997.

R. Stewart Johnston, *Scholar Gardens of China: A study and analysis of the spatial design of the Chinese private garden*, Cambridge University Press, Cambridge, 1991.

Maggie Keswick, 'The Gardens of China', *Design Book Review*, 1985, no. 7, pp. 38–40.

— 'Guide to Chinese Gardens', *Architectural Design*, 56, 9 (1986), pp. 41–47.

— 'An introduction to Chinese gardens', *The Authentic Garden: A Symposium on Gardens*, ed. L. Tjon Sie Fat & E. de Jong, The Clusius Foundation, Leiden, 1991, pp. 189–195.

Maggie Keswick, Judy Oberlander & Joe Wai, *In a Chinese Garden: The Art and Architecture of the Dr Sun Yat-sen Classical Chinese Garden*, The Dr Sun Yat-sen Garden Society of Vancouver, 1990.

Ellen Johnston Laing, 'Qiu Ying's Depiction of Sima Guang's Duluo Yuan and the View from the Chinese Garden', *Oriental Art*, vol. 33, no. 4 (winter 1987/88), pp. 375–380.

June Li & James Cahill, *Paintings of Zhi Garden by Zhang Hong: Revisiting a Seventeenth-Century Chinese Garden*, Los Angeles County Museum of Art, Los Angeles, 1996.

Liu Dunzhen, *Chinese Classical Gardens of Suzhou*, translated by Joseph C. Wang, McGraw-Hill, New York, 1993.

— 'The Traditional Gardens of Suzhou', translated by Frances Wood, *Garden History*, vol. 10, no. 2 (Autumn 1982), pp. 108–141.

John Makeham, 'The Confucian Role of Names in Traditional Chinese Gardens', *Studies in the History of Gardens & Designed Landscapes*, vol. 18, no. 3, Autumn 1998, pp. 187–210.

Stephen Markbreiter, 'Yu Yuan: A Shanghai Garden', *Arts of Asia*, vol. 9, no. 6, November–December 1979, pp. 99–110.

Georges Métailié, 'Insight into Chinese traditional botanical knowledge', *The Authentic Garden: A Symposium on Gardens*, ed. L. Tjon Sie Fat & E. de Jong, The Clusius Foundation, Leiden, 1991, pp. 215–223.

— 'Lettrés jardiniers en Chine ancienne', *Journ. d'Agric. Trad. et de Bota. Appl.*, nouvelle série, 1995, vol. 37, no. 1, pp. 31–44.

— 'Some Hints on 'Scholar Gardens' and Plants in Traditional China', *Studies in the History of Gardens and Designed Landscapes*, vol. 18, no. 3, Autumn 1998, pp. 248–256.

Georges Métailié & Nicole Staüble Tercier, *Au jardin potager chinois*, Alimentarium Vevey, 1997.

John Minford, 'The Chinese Garden: Death of a Symbol', *Studies in the History of Gardens & Designed Landscapes*, vol. 18, no. 3, Autumn 1998, pp. 257–268.

Kiyohiko Munakata, 'Mysterious Heavens and Chinese Classical Gardens', *Res 15* (Spring 1988), pp. 61–88.

— *Sacred Mountains in Chinese Art*, Krannert Art Museum & University of Illinois Press, Urbana & Chicago, 1991.

Michèle Pirazzoli-t'Serstevens, Anne Chayet et al., *Le Yuanmingyuan: jeux d'eau et palais européens du XVIIIe siecle à la cour de Chine*, Editions Recherche sur les civilisations, Paris, 1987.

Pierre & Susanne Rambach, *Jardins de longévité*, Skira, Geneva, 1987.

— *Gardens of Longevity in China and Japan: The Art of the Stone Raisers*, Rizzoli, New York, 1987.

Edward H. Schafer, 'Hunting Parks and Animal Preserves in Ancient China', *Journal of the Economic and Social History of the Orient*, no. 11, 1968, pp. 318–343.

David Ake Sensabaugh, 'Fragments of Mountain and Chunks of Stone: The Rock in the Chinese Garden', *Oriental Art*, vol. 44, no. 1, Spring 1998, pp. 18–27.

Osvald Sirén, *China and Gardens of Europe of the Eighteenth Century*, Ronald Press Co., New York, 1950.

— 'Architectural Elements of the Chinese Garden', *Architectural Review*, vol. 103, no. 618, June 1948, pp. 251–258.

Victoria M. Siu, 'China and Europe intertwined: a new view of the European sector of the Chang Chun Yuan', *Studies in the History of Gardens and Designed Landscapes*, vol. 19, no. 3/4, Autumn/Winter 1999, pp. 376–393.

Rolf A. Stein, *Le Monde en petit: Jardins en miniature et habitations dans la pensée religieuse d'Extrême-Orient*, Paris, 1987.

— *The World in Miniature: Container Gardens and Dwellings in Far Eastern Religious Thought*, translated by Phyllis Brooks, Stanford University Press, Stanford, 1990.

Jan Stuart, 'Ming dynasty gardens reconstructed in words and images', *Journal of Garden History*, vol. 10, no. 3, 1990, pp. 162–172.

— 'A Scholar's Garden in Ming China: Dream and Reality', *Asian Art*, vol. 3, no. 4, Fall 1990, pp. 31–51.

Régine Thiriez, *Barbarian Lens: Western Photographers of the Qianlong Emperor's European Palaces*, Gordon & Breach Publishers, Amsterdam, 1998.

Norah Titley & Frances Wood, *Oriental Gardens*, British Library, London, 1991.

L. Tjon Sie Fat & E. de Jong ed., *The Authentic Garden: A Symposium on Gardens*, The Clusius Foundation, Leiden, 1991.

Frances Ya-sing Tsu, *Landscape Design in Chinese Gardens*, New York, 1988.

Chuin Tung [Tong Jun], 'Foreign Influence in Chinese Architecture', *T'ien Hsia Monthly*, vol. VI, no. 5, May 1938, pp. 410–417.

Peter Valder, *The Garden Plants of China*, Weidenfeld & Nicolson, London, 1999.

— *Gardens in China*, Timber Press, Portland, Oregon, 2002.

L. Vandermeersch, 'L'arrangement de fleurs en Chine', *Arts Asiatiques*, vol. XI, 1965, fasc. 2, pp. 79–140.

Richard Vinograd, 'Family Properties: Personal Context and Cultural Pattern in Wang Meng's *Pien Mountains* of 1366', *Ars Orientalis*, vol. 13, 1982, pp. 1–29.

Joseph Cho Wang, *The Chinese Garden*, Oxford University Press, Hong Kong, 1998.

Wang Yi, 'Interior Display and its Relation to External Spaces in Tradition Chinese Gardens', translated by Bruce Doar & John Makeham, *Studies in the History of Gardens & Designed Landscapes*, vol. 18, no. 3, Autumn 1998, pp. 232–247.

Hellmut Wilhelm, 'Shih Ch'ung and his Chin-ku-yüan', *Monumenta Serica* 18, 1959, pp. 314–327.

Wong Young-tsu, *A paradise lost: the imperial garden Yuanming Yuan*, University of Hawai'i Press, Honolulu, 2001.

Frances Wood, 'The Traditional Gardens of Suzhou', *Garden History*, vol. 10, no. 2 (Autumn 1982), pp. 108–141.

Wu Shih-chang, 'Notes on the Origin of the Chinese Private Gardens', translated and condensed by Grace M. Boynton, *China Journal* 23, July 1935, pp. 17–22.

Xu Yinong, *The Chinese city in space and time: the development of urban form in Suzhou*, University of Hawai'i Press, Honolulu, 2000.

— 'Interplay of image and fact: the Pavilion of Surging Waves, Suzhou', *Studies in the History of Gardens & Designed Landscapes*, vol. 19, no. 3/4, Autumn/Winter 1999, pp. 288–301.

Yang Hongxun, *The Classical Gardens of China: History and Design Techniques*, translated by Wang Huimin, Van Nostrand Reinhold Co., New York, 1982.

Judith Zeitlin, 'The Secret Life of Rocks: Objects and Collectors in the Ming and Qing Imagination', *Orientations*, vol. 30, no. 5, May 1999.

Zhang Dai, *Souvenirs rêvés de Tao'an*, traduit du chinois, présenté et annoté par Brigitte Teboul-Wang, Connaissance de l'Orient, Gallimard, Paris, 1995.

Zhong Ming, 'An approach to the recording and preservation of Chinese gardens', *The Authentic Garden: A Symposium on Gardens*, ed. L. Tjon Sie Fat & E. de Jong, The Clusius Foundation, Leiden, 1991, pp. 225–230.

Zhu Junzhen, *Chinese Landscape Gardening*, Foreign Languages Press, Peking, 1992.

INDEX

Page numbers in italics refer to illustrations and captions

A
'Allegory Mountain' (Yu Shan Garden) 11
Allom, Thomas *213*
Amitabha's paradise 91, 214
Anemone japonica 191
Angler's Rest (Diaoyutai) 219
An Lushan rebellion 229
Anhui 175, 199, 228
apricot 39, 65, 209;
    Japanese apricot (*Prunus mume*) 192
Artless Administrator's Garden (Zhuo Zheng Yuan) 117, 121, *121*, 182
Attiret, Father 17–20, 21, 23, 37, 76, 153, 220
azalea, see rhododendron

B
Bai Juyi *95*, 95–6, 100, 101, 123, 181, 184, 187, 198, 227, 228
Bai Shi Ting (Pavilion for Obeisance to Rocks) 170, *170*
Bai Ta Shan (White Tower Hill) *71*
Ba Jin 14
bamboo 26, 32, 66, 68, 95, 96, 97, 100, 101, 118, 143, 146, 167, 187, 191, 192, 199, 203, 223
Bamboo Garden (Zhuyuan) 219
Ban Gu 50
Ban Mu *170*
Bazhou 199
begonia *157*, 196
Beihai Park/Gardens 14, 71, *163*, 184, 202
Benoit, Father *73*, 181, 220
Blue Waves Pavilion, see Cang Lang Ting
bonsai 229; see also miniaturisation
'Book of Change', see *Yi Jing*
'Book of Odes', 'Book of Songs', see *Shi Jing*
Book of Rites 197
Buddha, Buddhism, Buddhists 23, 47, 85, 91–2, 95, 130, 179, 187, 192, 194, 214, 224

C
Caesalpinia sepiaria 191
camellia 143, 189
Cang Lang Ting (Blue Waves Pavilion) *98*, 99, *148*, *163*, 221, 225
Canton 23, *149*, 165, 189, *212*
Cao Xueqin 219, 220

Caozhou 199
cassia 196
Catalpa bungei 195
Cedrela chinensis (red cedar) 194-5
Chambers, Sir William 23
Chang'an 59, *62*, 94, 194, 198, 228
Chang Chun Yuan
    (Garden of Expansive Spring) 72
Chen Congzhou, Prof. 226
Chengde (Jehol, Rehe) 8, 220-21
Chengdu 26, 229
Chen Haozi 100
Chen Hongshou 227
chrysanthemum 39, *92*, 93–4, *94*, 99, 101, 189, 193, 197, 199, 202–4, 210;
Chrysanthemum indicum 193
Chu Ci ('Songs of Chu') 42, 46, 197, 225
Cixi, Empress Dowager
    52, 71–2, 76, *78*, 82, 207, 220
clematis 189
Coleridge, Samuel Taylor 71, 114
Confucius, Confucianism 22, 23, 52–3, 54, 85, 87, 95, 96, 97, 131, 179, 185, 186, 192, 209, 210, 211, 214, 215, 221
Couple's Garden (Ou yuan) *12*, 224
crab apple 167, 189, 204
Crane Garden (He yuan) 226
Cui Xiu Tang *138*
cymbidium, see orchid
cypress 88, 190, 221

D
Dagoba 71, *51*
Da Guan Yuan ('Prospect Garden') *130*, 220
Daming Hu (Great Brightness Lake) 221
Daming Si (Temple of Great Enlightenment) 223
Dao 86-7, 88, 167, 184, 207, 209, 211, 214
Dao De Jing 85, 184,
Dao Ji 126, 127, *127*
daphne 189; *Daphne genkwa* 191
Da Shui Fa (Great Fountains) *73*
Dian Chun Yi *27*
Dipper Garden (Shaoyu) 11, 220
Dongguan 229
Dong Qichang 11, 226

Dong Zhongshu 52
Dowager Empress, see Cixi
Dream of the Red Chamber
    (*The Story of the Stone*) 14, 15, *40*, 129-30,
    130, 163-7, 196, 209-11, 219
Du Fu 229
Du Fu Caotang 229
Du Jin *210*
Du Le Yuan (Garden of Solitary Enjoyment)
    11, *96*, 96
Du Wan 176

E
'Eight Riders in Spring' *111*, 111-12
Elgin, Lord (James Bruce, 8th Earl of Elgin) 57
Emperor Ming Huang's Journey to Shu
    103, *103*, 105

F
Fan Chengda 193
fang (pavilion) 133
Fangzhang (Immortals' island) 50
Fan Kuan 112, *112*, *113*
fengshui (geomancy) 9, 67
Fenyang bieshu/Villa 227
Firmiana simplex, see *Sterculia platanifolia*
forsythia 189
Fortune, Robert 189-90
Foshan 229
Fountain Spring (Baotu quan) 221
Fouquet, Nicolas 54

G
Gao Fenghan *92*
Garden for Lodging One's Expansive Feelings
    (Jichang Yuan) plan *81*, 222
'Garden of Harmonious Interest'
    (Xie Qu Yuan) *78*, plan *80*, *82*, 220
Garden of Perfect Brightness
    (Yuan Ming Yuan)17
Garden of Solitary Enjoyment, see Du Le Yuan
Garden of the Master of the Fishing Nets,
    see Wang Shi Yuan
Genghis Khan 69
Gen Yue ('Immovable Peak') 67, 68, 218, 226
Ge Yuan 223
Ge Yuliang 226

Ginkgo biloba 143, *189*, 190
Golden Lotus 130
Goldsmith, Oliver 37
Gong, Prince 11, 219, 220
Gong Wangfu 219
Great Wall 70, 76, 146
Guangdong 11, *154*, 181, *181*, *203*, 207, 229
Guangxi *172*
Guangxu, Emperor 71-2
Guan Yun Feng (Cloud Crowned Peak) *176*
Guanzi 184
Gu Dashen 227
Guilin 39, 65, *172*
Guo Xi 114, *115*, 126
Guozhuang (Mansion of Guo family) 227
Guyiyuan ('Ancient Garden of Elegance') 226

H
'Hall of Incense to the Buddha' 78
'Hall of the Flower with the Green Calyx' 65
Hamamelis mollis (witch-hazel) 190
Han dynasty (206 BC–AD 220) 24, 45, 48, 50,
    51, 52, 54, 60, 64, 85, 169, 189, 193, 196, 197
Han Gaozu 51
Hangzhou 26, *65*, 68-9, 76, 96, 148, *178*, 181,
    186, 187,
Han Wudi (Martial Emperor)
    45, 50, 51, 59, 169, 218
Happy Sea 57
He Xiangu 192
He Yuan 223
hibiscus 42
Himalayas (Kunlun mountains) 39, 47, 190
honeysuckle 189
Hong Kong 6, 7, 48, 127, 229, 230
Hong Xue Yin Yuan Tu Ji 87, *149*, *152*, 170, *213*
Hou Hai *153*
Huang Chao 194
Huang Gongwang 125, 126
Huang He (Yellow River) 39
Huang Youzao 200
Huanxiu shanzhuang 225-6
Hua Qing hot spring *62*, *62*, 228
Hua Shan 46, *106*
Huiyuan 92
Huizong, Emperor 63-8, 70, *86*, 169, 226

I

Immortals, Immortality 45, 46-7, 48, 50-51, 67, 85, 103, 126, 127, 145, 169, 185
Imperial City, Peking 17, 21, 200
Island of Small Seas 178
Island of the Dragon King 76, 79
Islands/Isles of the Immortals 9, 50, 59, 64, 71, 170, 209

J

Jade Fountain spring/stream 76, 218, 220
Jade Grace Pavilion 144
Japan 48, 51, 82, 186;
    Japanese gardens 14, 212
jasmine 189
Jehol 76; see also Chengde
Jesuits 17, 76, 136;
    see also Attiret; Benoit; Ricci
Jiading 226
Jiangnan 6, 7, 11, 12, 85, 136
Jiangsu 118, 175, 136, 157
Jiangxi 12
Ji Chang Yuan (Garden for Lodging One's Expansive Feelings) 37, plan 81, 222
Ji Cheng 6, 9, 10, 11, 24, 115, 116, 222;
    see also Yuan Ye
Jie, king of Xia 58
Jinan 221
Jing Hao 106
Jing Shan (Coat Hill) 21
Jin Gu (Golden Valley) garden 89
Jin Shui (Golden Water) stream 20
Jixiao Shanzhuang 223
Johnston, Reginald 219
juniper 190

K

Kaifeng 64, 179
Kang Shen 219
Kangxi, Emperor 72, 186, 218, 220, 222
Kan Song Du Hua Xuan (Hall from which One Looks at the Pines and Contemplates the Paintings) 29, 32
Karakorum 70
Ke Yuan ('Could-be Garden') 229
Kinsai (Hangzhou) 186
Koelreuteria 196
Korea 59, 60,
Kubilai Khan 69-71, 218
Kuiji 90, 92
Kunlun mountains (Himalayas) 46, 47, 169
Kunming lake 57, 76, 78, 82, 181

L

Lake Tai 66, 67, 172, 181, 199, 222;
    see also Tai Hu
Lan Pu (Orchid Garden/Nursery) 165, 229
Lan Ting/Lanting (Orchid Pavilion) 90, 227
Lao Chou (Zhou) 207
Lao Zi 85
Leng Mei 133
Le Nôtre, André 17, 18, 136
Liangyuan 229
Li Bai 197, 227
Li Deyu 100, 101, 169
Liezi 45
Li Gefei 101
Li Gonglin 112, 113
lilac 78, 189, 191, 207; see also Syringa
lily 189, 191
Ling Yin Temple 186, 187
Lin Qing 87, 149, 152, 170, 213
Lion Grove (Shi Zi Lin) 117, 117, 121, 150, 153, 171, 176,178, 224
Li River 172
Li Sao 197
Li Shan 199
Li Song 204
Listening to the Wind 116, 116
Liu Bang 85, 88
Liu Meng 193
Liu Yongfeng 225
Liu Yuan ('Garden to Linger In') 48, 135, 140, 150, 159, 207, plan 215, 216, 225
Li Xian 42
Li Yu (Li Liweng) 124, 126, 147, 148, 170, 172, 174, 175, 196
Li Yuan 54, 66, 181, 222
Longevity Hill/Mountain (Wan Shou Shan) 52, 53, 57, 65, 76, 78, 79, 82, 220

Longhua Botanic Gardens (Shanghai) 48
Loropetalum chinensis 190, 191
Lorrain, Claude 211
lotus (Nelumbo speciosum) 42, 60, 95, 159, 184, 192, 197, 199, 200-202, 202, 207
Louis XIV (the Sun King) 22
Lou Lim-leok Garden 209, 230
Lu Hong 95
Lu Jin 224
Lüliang waterfall 87
Luo Ping 132
Luoyang 26, 46, 59, 101, 198, 199, 227
Lu Shan 92, 95, 184
Lu Xun 228
Lu You 199, 227

M

Macao 209
magnolia 26, 28, 143, 190
Ma Lin 116, 116
Manchus 17, 57, 58, 76, 184, 195, 220;
    see also Qing dynasty
Mao Zedong (Chairman Mao) 149, 165
Marble Boat (teahouse of Cixi) 76, 78, 132
Maries, Charles 190
Ma Yuan 53
Mei Yuan (Prunus garden) 66, 222
Mencius (philosopher) 58, 59, 72, 82
metasequoia 143
Mi Fu 113, 169-70
Ming dynasty (1368–1644) 8, 11, 12, 21, 40, 71, 96, 118-21, 126, 136, 137, 191, 218, 220, 221, 222
Ming Huang, Emperor 61-3, 70, 103, 196, 197, 228-9
Ming of Wei, Emperor 178
miniaturisation 48-50, 172
Mi Wanzhong 11, 220
Mochou Lake 130, 222
Mongols 69-70, 121
Mo Shilong 126
Mount Hui 222
Mount Li 228
Mount Lu (Lu Shan) 95
Mount Meru (Sumeru) 91
Mount Xi 222
mudan, see peony
mulberry 41, 53, 194
mushroom of immortality (lingzhi) 48, 67

N

Nan Hai (Southern Sea) 71, 218
Nanking (Nanjing) 26, 152,153, 204, 221-2
Nanxiang 226
Nanxun 226
Nanyang 197
Needham, Joseph 61
Nei Yuan (Inner Garden) 132, 141, plan 142, 144-5, 145
Nelson Wu 209, 216
Nicholson, Francis 24
Ningpo 167, 189, 228
Niu Sengru 169
Ni Zan 117, 117, 118, 118, 121, 125,

O

oleander 140
orchid 42, 191, 192, 199, 204, 229;
    cymbidium 192, 204, 229
Orchid Pavilion 90
osmanthus 196, 222
Ouyang Xiu 223
Ou Yuan (Couple's Garden) 224

P

Pan En 140, 141, 226
Pan Gu 184
Panyu 229
Pan Yunduan 226
paulownia 195
Pavilion for Obeisance to Rocks (Bai Shi Ting) 170, 170
Pavilion of Joy in Agriculture 53
'Pavilion of the Accumulated Void' 32
peach 39, 60 , 63, 189, 195, 198;
    Prunus persica 189
'Peach Blossom Spring' 24
pear 39, 189, 195
Peking 11, 12, 14, 15, 17, 19, plan 20, 21, 26, 57, 69, 76, 136, 150, 153, 159, 170, 184, 189, 195, 200, 204, 207

Penglai (Immortals' island) 50
peony 89, 199, 200, 202, 219, 224;
    tree peony (mudan)189, 197-9
persimmon 39, 195
Pianshi Shanfang 223
pine (Pinus) 66, 95, 100, 101, 116, 167, 172, 190, 191, 192, 196, 199, 221;
    Pinus bungeana 190
Ping Quan (Level Spring) 100
Pingshantang (Hall Level with the Mountains) 223
pinyin system 12
plum (Prunus) 17, 39, 60, 63, 65, 132, 146, 167, 193
Polo, Marco 68, 69, 70, 186, 189, 218
pomegranate 196
Pope, Alexander 23, 136,
poplar 196
Portugal, the Portuguese 189, 230
Poussin, Nicolas 211
primrose, primula 189, 191;
    Primula obconica 190
Prunus (plum) 192, 193, 196; P. mume (Japanese apricot) 192; P. persica (peach)189
Pu Ru (Pu Xinyu) 219
Pu Songling's House 221
Puyi, Emperor 219

Q

qi ('vital spirit') 50, 86, 106, 209; qi yun sheng dong 106, 111, 126, 127, 135, 210
Qi 103
Qian Gu 90
Qianlong Emperor 17, 19, 32, 37, 52, 72, 76, 78, 80, 136, 160, 163,181, 191, 192, 204, 218-9, 220, 221, 222, 223
Qianxian county 42
Qianzhang Palace 50
Qi Biaojia 11, 129, 140
Qin dynasty (221–207 BC) 44, 50,
Qing chengshan (Green Fortification Mountain) 229
Qing dynasty (1644–1911) 10, 26, 40, 71, 72, 126, 186, 193, 195, 218, 221, 222, 223,
Qinghui yuan 229
Qin Shi Huangdi 44-5, 50
Qionghua Island 71
Qiu Jin 228
Qiuxia pu (Garden of Autumn Vapours) 226
Qiu Ying 15,89, 96, 96, 103,106 123-4, 125, 176
Qufu 221
Qunxing caotang 229

R

Rehe, see Chengde
rhododendron (azalea) 37, 189
Ricci, Matteo 219
rose 26, 147, 189, 191
Royal Horticultural Society 189
Ruan Dacheng 222

S

Sanwei shuwu (Three Flowers Studio) 228
Sargentodoxa cuneata 191
Scholars' Pavilions. . . 112, 112
'Sea Palace parks' 41, 71, 76
Seven Sages of the Bamboo Grove 85, 88
Seven Star Crags 181, 181
Seventeen Arch bridge 57, 76, 160
Shaanxi 2, 42, 61, 62, 106
shamans 42, 50
Shandong 42, 199, 207
Shang Di 39
Shangdu (Xanadu) 71
Shang dynasty (c. 1600–1038 BC) 39, 58
Shanghai 48, 51, 86, 135, 137, 141, 169, 175, 185, 190, 199
Shanglin Park 45
shan shui (landscape) 170
Shanxi 2, 227
Shao, Duke of 195
Shaoxing 227
Shaoxing (Dipper Garden) 11, 220
Shen Bingcheng 224
Shen Fu 99, 197, 203, 204, 213-14
Shenstone, William 24
Shen Wu Men (northern gate) 21
Shen Yuan (garden) 227
Shen Yuan (painter) 19, 32, 196,

Shexian 228
Shi Chong, Duke 88, 89, 123
Shi Jing ('Book of Odes', 'Book of Songs') 41, 43, 51, 191, 193, 194, 195, 197
Shi Zhengzhi 193
Shi Zi Lin (Lion Grove) 117, 117, 121, 150, 153, 171, 176,178, 224
Shou Xihu (Slender West Lake) 223
Shun, Emperor 52
Sichuan 37, 199, 207
Sima Guang 11, 96-7, 100, 101, 123, 176
Sima Xiangru 45
Singapore 127
Sitwell, Osbert 204
Six Chapters of a Floating Life 97-9, 197, 203
Song dynasty (960–1126) 8, 26, 63, 69, 96, 114, 125, 136, 137, 174, 179, 192, 193, 199
Songjiang 11, 226
Song Qingling 220
Song Shangfu 199
Sophora japonica 30, 196,
spirea 189
Sterculia platanifolia (Firmiana simplex) 195
Stourhead 24, 200
Su Causeway 221
Su Dongpo (Su Shi) 96, 100, 113, 114, 117, 178, 181, 192, 227
Su Hua (Ling Shuhua) 204-7
Sui dynasty (589–618) 59-60, 223
Sui Yangdi 59-60, 181
Sui Yuan 152, 153
Sumeru (Mount Meru) 91
Summer Palace 14, 52, 57, 76, 82, 82, 150, 155, 159, 181, 207, see also Yi He Yuan
Sun Kehong 88
Sun King (Louis XIV) 54
Sun Yat-sen, Dr (Sun Zhongshan) 219, 220
Su Shunqin (Su Zimei) 225
Suzhou 7, 8, 9, 10, 11, 14, 53, 67, 82, 96, 99, 117, 118, 159, 163, 172, 178, 182, 186, 193, 199, 200, 203, 216 MORE
Switzer, Stephen 23
Symplocos paniculata 191
Syringa julianae 191

T

Tai Hu (Lake Tai) 30, 176
Taihu stones/rocks 86, 112, 124, 126, 141, 170, 176, 225
Taiping rebellion 141, 222, 224, 225
Tai Shan (Mount Tai) 46
Taiwan 209, 219, 221
Tang Dai 19, 32, 196
Tang dynasty (618–907) 8, 15, 39, 46, 60, 61, 63, 64, 94, 106, 169, 174, 184, 185, 192, 193, 194, 196, 197, 223
Taoists, Taoism 22, 23, 47-8, 82, 85-8, 106, 118, 172, 175, 184, 190, 195, 207, 211, 214, 215
Tao Qian (Tao Yuanming) 90, 92-4, 94, 96
Tea House (of Dowager Empress Cixi) 76
Temple, Sir William 23
thuja 196
Tian An Men Gate 219
Tianjin (Tientsin) 189, 219
Tianmu Mountain 178
Tianpeng 199
Tianru 224
Tianyige Library 167, 228
Tieshanyuan (Iron Hill Garden) 221
Tie Xuan 221
Tiger Balm gardens 127
Tiger Mountain 172, 226
ting (pavilion) 132-3, 147, 153, 200
Tong Jun 136, 144, 152
Tongli 157
Tui Si Yuan (Retreat to Think Garden) 157

U

Unsuccessful Politician's Garden, see Zhuo Zheng Yuan

V

Vaux-le-Vicomte 54
Versailles 17, 17, 21, 22, 136, 217
'Villa of the Golden Valley' (Jin Gu) 88, 89

W

Waley, Arthur 95
Wang Chuan 104, 105, 106
Wang family 224
Wang Meng 126

Wan-go Weng 105
Wang River 94, 105
Wang Shi Yuan (Garden of the Master of the Fishing-nets) 10, *15*, 24-37, *plan 25, 41*, 54, *147, 157, 163*, 224
Wang Tingna 228
Wang Wei 94-5, 103, 104-6, 111, 122, 123, 141, 209
Wang Xianchen 121, 224
Wang Xizhi 90, 227
Wang Yongkang 224
Wan Shou Shan (Longevity Hill) *53, 57*
weigela 189
Weize 178
Wu Kuan 225
Wen, king of Zhou 41, 58, 59
Wen Zhengming 117, 118–22, *122*, 123, *123*, 224, 225
Western Garden 141
Western Hills 70, 76, 79, *82*
West Lake (Xi Hu) 68, 116, 148, *178*,181, 186-7, 227
Wilde, Oscar 210
willow 17, 41, 132, 143, 186, 194, 199, 221
Wilson, Ernest 190-91
wintergreen 190
wisteria 150, 155, 191, 207
witch-hazel (*Hamamelis mollis*) 190
Wordsworth, William 114
Wu Daozi 106
Wu Sangui, General 224
wu wei 87
Wuxi 26, *37, 54, 66, 78, 80, 135, 176*, 222
Wu Zetian, Empress 61, 64

X
Xanadu (Shangdu) 71
Xia dynasty (?–2500 BC) 58
Xi'an (Chang'an) 62
Xianfeng Emperor 58
Xia Yuan 221
Xiao Canglang 221
Xiao Fei Hong (Flying Rainbow Bridge) *123*
Xiaolianzhuang (Lesser Lotus Manor) 226
xie (pavilion) 133
Xie clan 90
Xie He 106
Xie Kun 90
Xie Lingyun 90, 91, 92, 93, 181, 197
Xie QuYuan ('Garden of Harmonious Interest') *37, plan 80, 82*, 220, 222
Xi Hu (West Lake) *68*, 227
Xin'an Beiyuan 228
Xi Shang pavilion 219
Xiuning 228
Xi Wang Mu *106*, 195
Xi Yuan (Western Garden) *135*, 225
Xuan, king of Qi 58
Xu Da 221
Xu Garden (Xu Yuan) 222, 223
Xu Shitai 225
Xu Wei 227
Xu Xi *204*

Y
Yang Guifei 61-2, *62*, 70, 197, 228
Yang Hongxun 181
Yangshuo *39, 172*
Yangtze River 60, 136, 185, 190, 224
Yangzhou *23*, 26, 60, *125*, 223

Yan Yu Lou (Tower of the Misty Rain) 186
Yao, Emperor 52
Yellow River, see Huang He
Yichang 190
Yi He Yuan (Summer Palace) *37, 53, 57, 57, 76*, *plan 78, 82, 150, 155, 160*, 220, 221
Yi Jing ('Book of Change') 42, 228
yin and *yang* 145, 170, 181, 185, 213
Ying Tai (Ocean Terrace) 71
Yingzhou (Immortals' island) 50
Yi Yuan ('Garden of Pleasure') 225
Yongle Emperor 71
You Qiu *195*
Yuan dynasty (1280–1368) 116, 121, 179
Yuan Guanghan 53-4, 85, 100, 169,
Yuan Mei *152*, 153, 195, 224
Yuan Ming Yuan ('Old Summer Palace','Garden of Perfect Brightness') 10, 17, *plan 18, 19*, 26, 57, 58, 60, 72, *73*, 76, 79, 153, 181, 184, 219, 220
Yuan Ye 6, 9, 10, 14, 23, 24, 111, 114, 115, 124, 125, 126, 135, 136, 138, 146, 147, 150, 153, 172, 175, 181, 210; see also Ji Cheng
Yuexiu Park *165*
Yu Hua Yuan (back garden of Imperial City) *48, 58, 76*, 112, *137, 157, 171, 207*, 219
Yun (wife of Shen Fu) 99, 197, 203
Yunnan 37
Yu Shan Garden ('Allegory Mountain') 11
Yuyin shanfang (Mountain Cottage of Abundant Shade) 229
Yu Yuan ('Garden to Please') *86, 130, 135*, 137-45, *141, plan 142, 143, 145, 146, 150*, 160, *169*, 226

Z
zhai (pavilion)133, 153
Zhang Heng 43
Zhang Hong *85, 198*
Zhang Lian 10, 11
Zhang Lun 169
Zhang Nanyang 226
Zhang Shi 222
Zhangwu zhi (*Treatise on Superfluous Things*) 225
Zhang Yanyuan 106
Zhan Yuan (Outlook Garden) 221, 222
Zhao Mengfu 113
Zhu De, Marshal *165*
Zhejiang 136, *167*, 228
Zhenghua *192*
Zhengzhou 46
Zhen Zhijie 61
Zhi Garden *85, 198*
Zhongshan Park *51*,159, 219
Zhou period (1066–221 BC) 42-3, 58
Zhou Dunyi 192
Zhuangzi 85, 87, 216, 218
Zhu Da (Bada shanren) 223
Zhu Mian 67-8
Zhuo Zheng Yuan (Artless Administrator's Garden, Unsuccessful Politician's Garden) 53, *105*, 117, 121, *121*, 122, 123, *129, 148*, 182, *plan 183*, 200, 224
Zhu Sansong 226
Zibo 221
Zixiu 68
Zuibaichi (Drunken Bai Pond) 226
Zuoyinyuan (Garden of Sitting in Reclusion) 228

ACKNOWLEDGMENTS

PHOTOGRAPHIC ACKNOWLEDGMENTS

With the exception of those listed below, all the gardens in this book were photographed by Maggie Keswick. For permission to reproduce archive material and for supplying the photographs on the following pages of this revised edition, the Publishers would like to thank:

Ingrid Booz Morejohn 5, 13, 16, 21, 29, 33, 38, 39, 49 above left, 62, 73 below, 79 above, 107 below left, 112 right, 122, 128, 131 above, 139 below, 140, 156 above left, 156 below left, 156 below right, 158 below, 166, 168, 173, 178, 184, 187, 188, 190, 199, 202
Bridgeman Art Library 212 (Private Collection)
©**British Museum** 56 (OA 1948.5.8.03). 63 (OA 1926 4-10-01), 104 (OA 1889 11-11-1)
Ann J. D. Griffin 30, 40, 83 below, 86, 143, 144 above, 201 below
Peter Hayden 1, 4, 6, 14–15, 27 below, 28, 36 above, 49 below left, 49 below right, 66, 79 below, 99, 105 right, 134 above left, 148 right, 156 above right, 156 centre left, 161, 179, 180 below, 189, 191, 218
Charles Jencks 59, 70, 82, 83 above, 174, 181, 208
Museum für Ostasiatische Kunst, SMPK, Berlin 84 (photo Stüning/1991), 198 (photo Stüning/1991)
Collection of the National Palace Museum, Taipei, Republic of China 44, 50 below, 64, 95, 102, 110, 112 left, 113 left, 113 right, 115, 116, 118, 133, 169, 211
The Nelson-Atkins Museum of Art, Kansas City, Missouri 51 (Purchase: Nelson Trust, 33-1482 A), 52–3 below (Purchase: Nelson Trust, 63-19), 107 right (Purchase: Nelson Trust, 47-71), 108–109 (Gift

of the Herman R. and Helen Sutherland Foundation Fund, F72-39), 194–5 (Purchase: Nelson Trust, 50-23)
Osaka Municipal Museum of Art, Abe Collection 93

Photographic sources for the original edition of this book:
Aerofilms Ltd. 17
Bibliothèque Nationale, Paris 19, 22, 34–5, 73 above, 197
©**British Museum** 24 above, 193
Avery Brundage Collection, San Francisco 88
Chion-in, Kyoto 89
Freer Gallery, Washington 47, 132 below right
Harvard Yenching Library 87, 170, 213
Honolulu Academy of Arts 94
Morse Collection, New York 121, 126
Collection of the National Palace Museum, Taipei, Republic of China 204
Private Collection 124–5, 216, 217
Collection Mr and Mrs C. C. Wang, New York 127, 175
Collection Wango H. C. Weng 96–7
Warren King Collection, Hong Kong 40

PUBLISHER'S ACKNOWLEDGMENTS FOR THIS REVISED EDITION
Project Editor Jo Christian
Designer Becky Clarke
Picture Editor Sue Gladstone
Production Kim Oliver